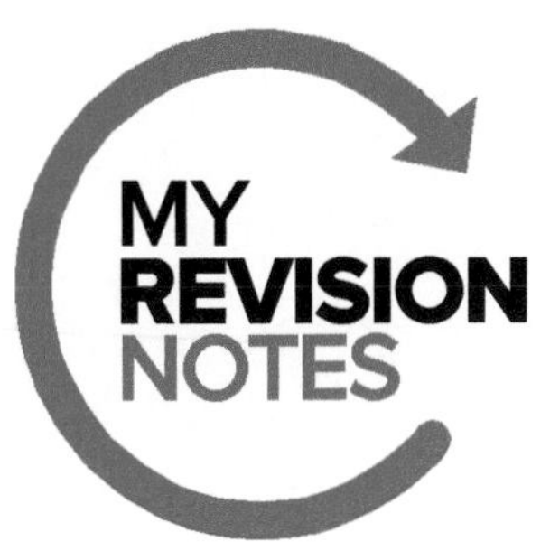

City & Guilds

Level 2 Technical Certificate (8202-25)

PLUMBING

Stephen Lane

HODDER EDUCATION
AN HACHETTE UK COMPANY

Orders: please contact Hachette UK Distribution, Hely Hutchinson Centre, Milton Road, Didcot, Oxfordshire, OX11 7HH. Telephone: +44 (0)1235 827827. Email education@hachette.co.uk Lines are open from 9 a.m. to 5 p.m., Monday to Friday. You can also order through our website: www.hoddereducation.co.uk

ISBN: 978 1 3983 2735 1

First published in 2021 by
Hodder Education,
An Hachette UK Company
Carmelite House
50 Victoria Embankment
London EC4Y 0DZ

www.hoddereducation.co.uk

Impression number 10 9 8 7 6 5 4 3

Year 2025 2024 2023

Cover photo © stuartbur – stock.adobe.com

Typeset in India.

Printed and bound by CPI Group (UK) Ltd, Croydon, CR0 4YY

A catalogue record for this title is available from the British Library.

Get the most from this book

Everyone has to decide his or her own revision strategy, but it is essential to review your work, learn it and test your understanding. These Revision Notes will help you to do that in a planned way, topic by topic. Use this book as the cornerstone of your revision and don't hesitate to write in it – personalise your notes and check your progress by ticking off each section as you revise.

Tick to track your progress 

Use the revision planner on pages 4–7 to plan your revision, topic by topic. Tick each box when you have:

+ revised and understood a topic
+ tested yourself
+ practised the exam questions and gone online to check your answers.

You can also keep track of your revision by ticking off each topic heading in the book. You may find it helpful to add your own notes as you work through each topic.

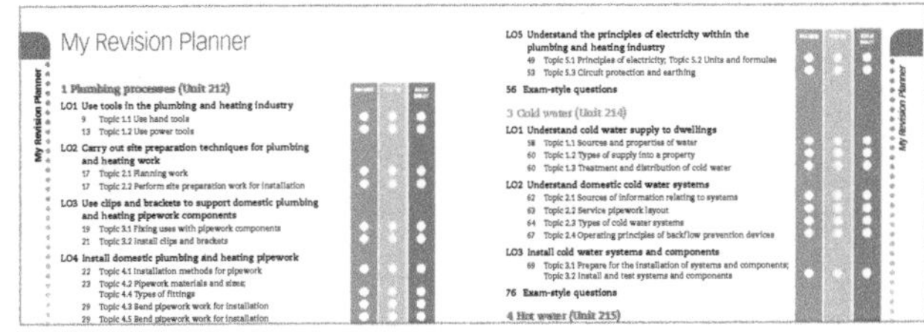

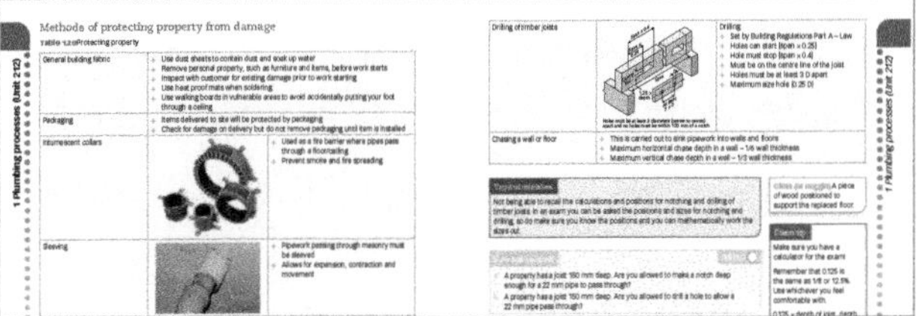

Features to help you succeed

Exam tips

Expert tips are given throughout the book to help you polish your exam technique in order to maximise your chances in the exam.

Typical mistakes

The author identifies the typical mistakes that candidates make in exams and explains how you can avoid them.

Now test yourself

These short, knowledge-based questions provide the first step in testing your learning. Answers are available online.

Definitions and key words

Clear, concise definitions of essential key terms are provided where they first appear.

Exam-style questions

Practice exam questions are provided for each topic. Use them to consolidate your revision and practise your exam skills.

Online

Go online to check your answers to the exam questions at **www.hoddereducation.co.uk/myrevisionnotesdownloads**

Check your understanding

These questions test your basic understanding of the information as you work through the course. Answers are available online.

Exam breakdown

For guidance on how you will be assessed and how to prepare for your exam, see the end of this book (page 162).

My Revision Planner

1 Plumbing processes (Unit 212)

REVISED | TESTED | EXAM READY

2 Electrical and scientific principles (Unit 213)

REVISED | TESTED | EXAM READY

REVISED | TESTED | EXAM READY

7 Health and safety and industry practices (Unit 211)

REVISED | TESTED | EXAM READY

Countdown to my exams

6–8 weeks to go

- Start by looking at the specification — make sure you know exactly what material you need to revise and the style of the examination. Use the revision planner on pages 4–7 to familiarise yourself with the topics.
- Organise your notes, making sure you have covered everything on the specification. The revision planner will help you to group your notes into topics.
- Work out a realistic revision plan that will allow you time for relaxation. Set aside days and times for all the subjects that you need to study and stick to your timetable.
- Set yourself sensible targets. Break your revision down into focused sessions of around 40 minutes, divided by breaks. These Revision Notes organise the basic facts into short, memorable sections to make revising easier.

REVISED

2–6 weeks to go

- Read through the relevant sections of this book and refer to the exam tips, summaries, typical mistakes and key terms. Tick off the topics as you feel confident about them. Highlight those topics you find difficult and look at them again in detail.
- Test your understanding of each topic by working through the 'Now test yourself' questions in the book. Look up the answers online.
- Make a note of any problem areas as you revise, and ask your teacher to go over these in class.
- Look at past papers. They are one of the best ways to revise and practise your exam skills. Write or prepare planned answers to the exam practice questions provided in this book. Check your answers online at **www.hoddereducation.co.uk/myrevisionnotesdownloads**
- Track your progress using the revision planner and give yourself a reward when you have achieved your target.

One week to go

- Try to fit in at least one more timed practice of an entire past paper and seek feedback from your teacher, comparing your work closely with the mark scheme.
- Check the revision planner to make sure you haven't missed out any topics. Brush up on any areas of difficulty by talking them over with a friend or getting help from your teacher.
- Attend any revision classes put on by your teacher. Remember, he or she is an expert at preparing people for examinations.

REVISED 

The day before the examination

- Flick through these Revision Notes for useful reminders, for example the exam tips, typical mistakes and key terms.
- Check the time and place of your examination.
- Make sure you have everything you need – extra pens and pencils, tissues, a watch, bottled water, sweets.
- Allow some time to relax and have an early night to ensure you are fresh and alert for the examinations.

REVISED 

My exams

825 Employer involvement

Hours completed............ (✔)

026 Synoptic test

Practical assessment

Date: ....

Time: ....

Location: ....

025/525 End of year exam

Multiple choice

Date:..

Time: ....

Location: ....

1 Plumbing processes (Unit 212)

This unit is mainly about identification of tools and fittings, use and maintenance. So, ask yourself these questions as you go through this unit:

+ What tools are used?
+ What types of building fabrics are found in a property?
+ Why is fitting selection important?

This unit holds a high weighting within the test specification, so you will need to get to grips with the content of this area.

LO1 Use tools in the plumbing and heating industry

Topic 1.1 Use hand tools

REVISED 

There are many different hand tools used in the plumbing and heating industry. You need to know the function, safe use and maintenance of these hand tools.

Table 1.1 Screwdrivers

Pozi drive		Used for tightening and loosening screws
Philips		
Flat head		
Safety	Be careful of the pointed end; make sure handle is not damaged	
Maintenance	Check handle is secure and drive end is not damaged	

Exam tip

When looking at the tools, make sure you know:

+ the name
+ what it does (its function)
+ safe use
+ any maintenance required.

So, remember when you use tools in the workshop or on site and what you used them for, as this will prepare you for the questions in the exam.

Table 1.2 Hammers

Claw		Driving in and removing nails
Club		Heavy hammering, used with chisels
Safety	Don't drop heavy items; wear eye protection	
Maintenance	Check handle and head are attached securely	

Table 1.3 Chisels

Bolster		Cutting brickwork and lifting floorboards
Flat		Cutting, breaking and chasing brickwork
Safety	Remove any 'mushroom head' from the tool. Mind your hand when hitting the chisel with the hammer	
Wood		Shaping wood and notching floorboards
Safety	It has a very sharp cutting edge; wear eye protection	
Maintenance	Sharpen cutting edge when required	

Table 1.4 Grips and wrenches

Water pump pliers		General purpose grips
Pipe wrench – footprint type		General purpose LCS grips
Pipe wrench – stilson type		Installing LCS pipework
Basin wrench		Tightening or loosening hard to reach connections (e.g. basin and bath taps)
Mole grips		Sprung loaded grips
Safety	Don't catch fingers in the jaws; watch out for slipping off components	
Maintenance	Clean jaws and lubricate moving parts	

LCS (Low Carbon Steel)
Used for commercial pipework.

Check your understanding

1 Which tool would you need to remove a mushroom head from?
2 When installing an LCS pipework system, what tool would you use to grip the pipe with to connect it to a fitting?

Typical mistake

Never use footprint- or stilson-type pipe wrenches on copper or brass items, because the teeth on these tools will damage softer materials.

Table 1.5 Spanners

Adjustable		Tightening and loosening compression fittings and valves of various sizes
Open ended		Tightening and loosening set-sized fittings
Ring		Tightening and loosening set-sized fittings
Box		Tightening and loosening taps to sanitary ware
Immersion		Tightening and loosening immersion heater to a hot water cylinder
Safety	Adjust correctly or use the correct size to avoid slipping off or 'rounding-off' components	
Maintenance	Lubricate moving parts	

Exam tip

You might be asked how to size an adjustable spanner correctly. Always size an adjustable spanner correctly to avoid 'rounding off' the flat edges of a compression fitting and also to avoid hitting your hand if the tool slips off.

Table 1.6 Spirit levels

Torpedo or boat		Levelling smaller items, has magnetic strip which can help with radiators or boilers
Spirit level		Horizontal or vertical lines or leveling larger items like baths
Safety	Do not leave on the floor as this is a potential trip hazard	
Maintenance	Do not drop as that will affect the glass bubble	

Table 1.7 Manual pipe threader

Hand dies		Threading LCS pipework
Safety	Sharp cutting edge; swarf produced is sharp; lubricating oil used can cause a slip hazard or dermatitis	
Maintenance	Sharpen cutting edge or replace dies; lubricate rachet	

Table 1.8 Pipe cutters

Pipe slice		Cutting copper pipe (single size only)
Adjustable pipe cutter		Cutting copper pipe (within a range)
Plastic pipe cutter		Cutting plastic pipe (within a range)
Safety	Sharp cutting wheels or edges	
Maintenance	Replace blade or replace tool; lubricate moving parts	

Table 1.9 Saws

Hacksaw		Used to cut various materials to length according to the blade of the saw
Junior hacksaw		
Universal hard point saw		
Floorboard saw		
Pad saw		
Safety	Sharp teeth on cutting edge	
Maintenance	Replace blade if possible – teeth facing forward	

Exam tip

In an exam you might be asked which stroke or movement of the saw cuts the material. The teeth of a saw face forward, so the cut is made on the **forward** stroke.

Check your understanding

3 You are using a copper pipe slice. It is not cutting a single groove but tram lining down the pipe. What needs to happen to the pipe slice to correct this?

Table 1.10 Pliers

General purpose pliers		Grip and tighten items
Long nose pliers		Grip small items or reach into small places
Circlip pliers		Internal and external types Removing and replacing circlips on maintenance jobs
Side cutters		Cutting wire or cable to length
Safety	Jaws can pinch fingers	
Maintenance	Lubricate hinge and clean jaws	

Table 1.11 Bending tools

Scissor type bender		Forming copper pipe (set sizes only)
Micro-bore bender		Forming copper pipe (within a range)
Safety	Take care not to trap fingers	
Maintenance	Lubricate moving parts	

Exam tip

In the exam, you might be asked about the scissor type bender and its component parts.

- The guide or slip prevents the copper pipe rippling.
- The copper pipe goes in between the former and the guide, then the roller goes on top of the guide (as seen in the picture).

Now test yourself TESTED

1 You are replacing a central heating circulator and you tighten up the electrical connections in the terminal strip. What is the main difference between the electrical screwdriver you are using and a conventional screwdriver?

Topic 1.2 Use power tools

REVISED

Like with hand tools, there are many different power tools used in the industry, so you will need to know their function and how to use them safely.

Table 1.12 Power drills

Rotary hammer drill		Used to drill holes in building fabrics Has a standard self-centring chuck, variable speed, drilling and hammer drilling (Available in cordless)
SDS hammer drill		Powerful drill used to drill holes in building fabric, core drill and chase Has a bayonet-type (SDS) chuck, variable speed, drilling, hammer drilling and chuck lock (Available in cordless)
Safety	Electric shock, trip hazards, secure drill correctly Wear ear protection and other PPE as necessary	
Maintenance	PAT test, inspect, clean and lubricate chuck	
Cordless drill		Plumber's choice – flexible, fewer hazards Has a standard self-centring chuck, variable speed, drilling and hammer drilling Differing battery voltages and amp-hour charge (Available in SDS)
Safety	Inspection Wear eye protection and other PPE as necessary	
Maintenance	Inspect, clean and lubricate chuck. Re-charge battery	

Exam tip

Ask yourself why a cordless drill would be the tool of choice when drilling a hole in a customer's property.

Check your understanding

4 You need to drill an 8.0 mm hole in brickwork to install the support for a new boiler. Which power tool would you choose?

Typical mistakes

Not being able to recall the basic maintenance considerations when using tools. Remember that tools can get dirty, blunt, damaged or loose after use so it's important to look after them appropriately.

Table 1.13 Drill bits

Masonry drill bit		Tungsten tip to penetrate masonry
Wood drill bit		Point and two spurs to guide through wood
Metal drill bit		General purpose bit made of high-speed steel Can be used on wood, plastic and metal

Spade bit		Also known as a flat bit; used to drill larger holes Point to guide through soft wood
Core drill		Drilling large holes through masonry (soil/waste pipes and flues) Diamond tipped Do not use hammer action with these
Hole saw		Teeth are hardened Used to drill into cisterns or acrylic sanitary ware
Safety	Adjust correctly or use the correct size. Wear PPE as dust and particles are produced.	
Maintenance	Lubricate moving parts	

Table 1.14 Power saws

Circular saw		Used when lifting floorboards and notching joists
Jig saw		Used for cutting worktops for sinks and basins
Reciprocating saw		General purpose saw but not accurate
Safety	Inspection of tool, guard and triggers. Make sure blades are secure Wear eye protection and other PPE as necessary	
Maintenance	Clean, replace blades, check guard, PAT test and inspect	

Table 1.15 Portable pipe threading machines

Hand-held electric threader		Used to thread LCS pipework in situ
Pipe threading machine		Used to thread, cut and de-burr LCS pipework on site
Safety	PAT test, inspection, electric shock, swarf and oil Wear ear protection and other PPE as necessary	
Maintenance	PAT test, inspect, clean, sharpen dies and top up cutting oil	

Table 1.16 Hydraulic machine bender

Hydraulic machine bender		Used to form LCS pipework
Safety	High pressure oil used; care with hands and arms	
Maintenance	Check oil level	

Table 1.17 Hydraulic crimping tool

Hand-held crimping tool		Used to crimp or 'press fit' components in place Can be battery or mains powered
Safety	PAT test, inspection and electric shock Wear ear protection	
Maintenance	PAT test, inspect and clean jaws	

Table 1.18 Blow torch

Soldering torch		Used to solder copper pipe fittings
Safety	Flammable gas, burning customer's property and self Wear eye protection	
Maintenance	Clean jet and nozzle, replace gas cylinder when required	

Other specialised tools

- Fusion welder: Used by utility companies to connect mains water and gas pipework. Fittings have small electrical coils which heat up and melt items together.
- Freezing kit: Used to avoid full drain down of systems. Two sections of pipe are frozen and the pipework in-between is worked on.

Exam tip

In an exam it is important to remember that, when using a freezer kit, the water in the system pipework must not be allowed to flow to allow the water to freeze. Care must be taken because your hands can stick to the frozen clamps. So, a question could be associated with the use or health and safety when using a freezer kit.

Now test yourself

TESTED 

2 The blade in a power saw is blunt and requires replacing. What is the primary safety action to take before replacing the blade?

3 You are asked to assist in the installation of a new gas boiler in a customer's kitchen. A template identifies that a 100 mm hole is required to be put through the wall. What power tool and drill bit should be used?

LO2 Carry out site preparation techniques for plumbing and heating work

Topic 2.1 Planning work

REVISED 

There are different methods that you can use for planning work for installation. For example:

Table 1.19 Planning methods

Job schedules	+ Works programme is a time against activity chart + Includes start and finish dates + Outlines the order for work to be completed in
Materials list	+ Important list of all components required for the installation + Saves extra trips to the suppliers + Enables the formulation of an **estimate** or **quotation**
Storing tools and equipment	+ Prevent damage or theft + Time saving as you know where items are

Estimate An approximate price that could vary slightly.

Quotation A fixed price that cannot vary.

Topic 2.2 Perform site preparation work for installation

REVISED

In order to prepare work for installation on site, there are many different factors that must be taken into account.

Types of work environments:

- new site for new build properties
- refurbishing and existing property (empty or occupied)
- industrial or commercial properties (non-domestic)
- domestic properties.

Methods of protecting property from damage

Table 1.20 Protecting property

General building fabric	+ Use dust sheets to contain dust and soak up water + Remove personal property, such as furniture and items, before work starts + Inspect with customer for existing damage prior to work starting + Use heat proof mats when soldering + Use walking boards in vulnerable areas to avoid accidentally putting your foot through a ceiling	
Packaging	+ Items delivered to site will be protected by packaging + Check for damage on delivery but do not remove packaging until item is installed	
Intumescent collars		+ Used as a fire barrier where pipes pass through a floor/ceiling + Prevent smoke and fire spreading
Sleeving		+ Pipework passing through masonry must be sleeved + Allows for expansion, contraction and movement
Lifting timber floorboards	+ Tools: pencil, circular saw or floorboard saw, bolster chisel + Warn customer + Mark floorboards + Set circular saw to just under the depth of the floorboard + Cut length and cross + Lift using bolster chisel + Remove nails + A **cleat** (or **noggin**) may be required to support the floorboard or chipboard	

Notching of timber joists	Span × 0.25
	Span × 0.07
	0.125 × depth
	Span
		+ Notching:
		  + Set by Building Regulations Part A – Law
		  + Start of notch {span × 0.007}
		  + End of notch {span × 0.25}
		  + Depth of notch {0.125 × depth of joist}
		+ The span is the distance from supporting wall to supporting wall
		+ A nail guard or cover plate should be used over the joist before relaying the floorboard down

Drilling of timber joists	Span × 0.4 Span × 0.25 Centre line Span D 1.25 × depth At least 3 × D Holes must be at least 3 diameters (centre to centre) apart and no holes must be within 100 mm of a notch	Drilling: + Set by Building Regulations Part A – Law + Holes can start {span × 0.25} + Hole must stop {span × 0.4} + Must be on the centre line of the joist + Holes must be at least 3 D apart + Maximum size hole {0.25 D}
Chasing a wall or floor	+ This is carried out to sink pipework into walls and floors + Maximum horizontal chase depth in a wall – 1/6 wall thickness + Maximum vertical chase depth in a wall – 1/3 wall thickness	

Typical mistakes

Not being able to recall the calculations and positions for notching and drilling of timber joists. In an exam you can be asked the positions and sizes for notching and drilling, so do make sure you know the positions and you can mathematically work the sizes out.

Cleat (or noggin) A piece of wood positioned to support the replaced floor.

Now test yourself TESTED

4 A property has a joist 150 mm deep. Are you allowed to make a notch deep enough for a 22 mm pipe to pass through?

5 A property has a joist 150 mm deep. Are you allowed to drill a hole to allow a 22 mm pipe pass through?

Exam tip

Make sure you have a calculator for the exam!

Remember that 0.125 is the same as 1/8 or 12.5%. Use whichever you feel comfortable with.

0.125 × depth of joist, depth of joist / 8, depth of joist × 12.5/100

LO3 Use clips and brackets to support domestic plumbing and heating pipework components

Topic 3.1 Fixing uses with pipework components

REVISED

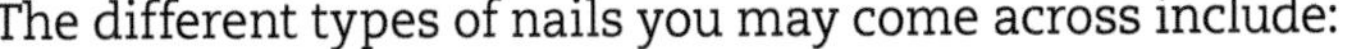

The different types of nails you may come across include:

+ **masonry nail**. Hardened steel. Used to make fixings in brickwork
+ **copper nail**. Used on sheet lead work to prevent corrosion
+ **round bright wire nail**. Used for general purpose woodwork
+ **oval bright wire nail**. Used for woodwork when appearance is important
+ **floorboard nail** (floor brads). Used to secure floorboards down.

The different types of screws you may come across include:

+ **countersunk screw**. Used for general purpose securing. Brass screws will have slotted head, steel will have pozi drive head. Can be coated to prevent corrosion. End up flush with surface
+ **raised countersunk screw**. Used for decorative fittings, made to be on show
+ **round head screw**. Used to secure copper saddles in place
+ **mirror screw**. Aesthetic chrome or plastic cap used to secure mirrors and bath panels in place
+ **coach screw**. Used to secure heavier items in place like boilers and larger radiators
+ **chipboard screw**. Deeper course thread used to secure chip and fibreboard.

Don't forget the screw head types covered earlier in Topic 1.1 under 'Screwdrivers': slotted, Phillips and pozi drive. Pozi drive is the most commonly used head type now, but you will come across slotted head screws where the screw is made of brass.

Screw materials:

- brass – expensive and corrosion resistant
- steel – general purpose; cheap but could corrode
- coated – general purpose; cheap and corrosion resistant
- stainless steel – more expensive, long-life and corrosion resistant.

Check your understanding

5 You have to secure a toilet pan in position. What material would the screws be made of and why?

6 You are installing a run of guttering around a property. What type of screw head would you use to secure the fascia brackets? What material would the screws be made of and why?

Exam tip

In an exam you might be asked why a countersunk screw is preferred or which type of screw head protects the customer from cutting their hands or feet. Countersunk screws are used to resecure floorboards back. The head sinks into the wood to give a flush finish.

Plastic plugs (rawlplugs)

There are many different ones which are used in conjunction with screws. The colours denote the sizes of rawlplugs but the two most common used by plumbers are:

Screw gauge 6–12

Drill diameter: 6.0 mm

Screw gauge 10–14

Drill diameter: 7.0 mm

Check your understanding

7 Where and why are rawlplugs used?

Exam tip

The 'Check your understanding' questions are very similar but the basis of these is two-fold – corrosion resistance and cost. In the first question, only a few screws would be used, so the more expensive corrosion resistant screws are used. For gutter installation, many screws would be used, so more economical corrosion resistant screws are used.

Other fixings

Table 1.21 Fixings

Coach bolt		These are not generally used by plumbers but can be found in building structures, holding up cold water cisterns in lofts
Anchor bolt		These are not generally used by plumbers but can be found connecting structure and non-structure items together
Rawlbolt		Used for securing very large size pipework into masonry. Sometimes known as an 'anchor bolt'

Toggle type fixing (Plasterboard fixing)		Spring-loaded butterfly clips used to hang radiators on plasterboard
Collapsing cavity fixing (Plasterboard fixing)		The strongest type of plasterboard fixing used to secure sanitary ware to plasterboard
Self-drill plasterboard fixing		The weakest plasterboard fixing used to secure small items to plasterboard
Drive in fixing		Used generally to secure guttering in place on older properties that do not have fascia boards

Check your understanding

8 Which plasterboard fixing would be used to secure items to a plasterboard wall in a newly installed bathroom?

Typical mistakes

Being unable to identify common fixings. Take another look at all the fixings and make sure you know their names and where they would be used.

Topic 3.2 Install clips and brackets

REVISED

Table 1.22 Clips and brackets

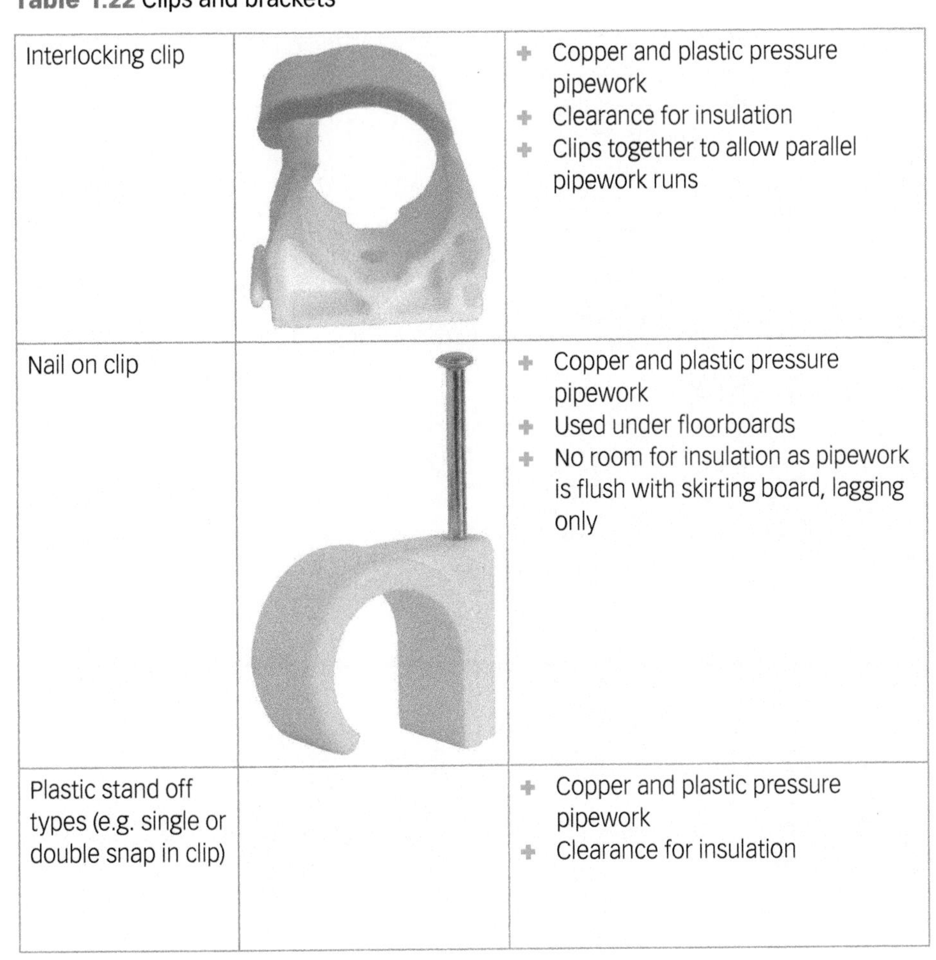

Interlocking clip		+ Copper and plastic pressure pipework + Clearance for insulation + Clips together to allow parallel pipework runs
Nail on clip		+ Copper and plastic pressure pipework + Used under floorboards + No room for insulation as pipework is flush with skirting board, lagging only
Plastic stand off types (e.g. single or double snap in clip)		+ Copper and plastic pressure pipework + Clearance for insulation

Copper saddle		+ Copper pipework + Used on skirting board + No room for insulation as pipework is flush with skirting board
School board clip or bracket (pressed brass)		+ Brass strip for copper pipework + Unscrews sideways + Used on walls and skirting boards
School board clip or bracket (cast brass)		+ Cast iron for LCS pipework + Unscrews sideways + Used on walls and skirting boards
Munsen ring and back plate (cast brass)		+ Cast brass for copper pipework + Unscrews top to bottom + Used on ceilings + Threaded bar screwed into the back (or base) plate allows height adjustment
Munsen ring and back plate (cast iron)		+ Cast iron for LCS pipework + Unscrews top to bottom + Used on ceilings + Threaded bar, screwed into the back plate and ring, allows height adjustment
Munsen ring with rubber lining		+ For copper and LCS pipework + Used to prevent any transfer of noise and vibration

Exam tip

A question could ask you about the clip material to support a type of pipe. **Never** use brass clips for LCS pipework and **never** use cast iron clips for copper pipework, as this will create galvanic corrosion. So, it should be copper pipe: brass clip and LCS pipe: cast iron clip.

LO4 Install domestic plumbing and heating pipework

Topic 4.1 Installation methods for pipework

REVISED

This is a workshop activity in which you will need to show that you know the methods used to install pipework, including the use of the following:

- prefabrication of pipework
- installation of pipework in-situ
- use of sleeves
- fire stopping to pipework.

Topic 4.2 Pipework materials and sizes; Topic 4.4 Types of fittings

REVISED

Pipework and their associated fittings will be covered here to help with your revision.

Table 1.23 Pipework and fittings

Copper – BS EN 1057	
R220 soft coils	+ Soft tube – fully annealed, can be carefully bent by hand + 6, 8, 10 mm microbore central heating + 15, 22, 28 mm small bore underground water services + Supplied in coils up to 50 m in length
R250 half hard lengths	+ Half hard tube – bent in scissor-type benders + 15, 22, 28 mm small bore + 35, 42, 54 mm large bore + Supplied in straight lengths of 3 or 6 m
R290 hard lengths	+ Hard tube, thin walled + No longer used

Exam tip

In an exam you might be asked which class and size of copper pipe is used in certain installations. The most common tubes used in domestic central heating, cold and hot water systems, are 15 and 22 mm R250 (half hard copper). Whereas 8, 10 and 12 mm R220 (soft copper) are used in microbore central heating systems.

Copper pipe fittings

Table 1.24 Copper tube fittings recognition

	Couplers	Equal tees (all three connections equal size)	Elbows and bends	Reducers
End feed			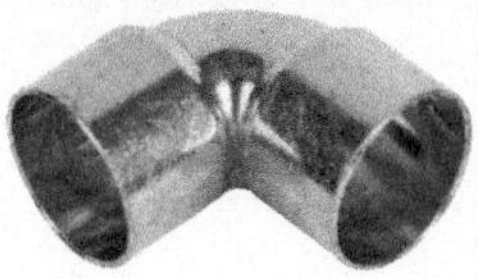	
Integral solder ring				
Compression				
Push fit				

Table 1.25 Reducing tees

Reduced end	Reduced branch	Reduced end and branch	Two reduced ends
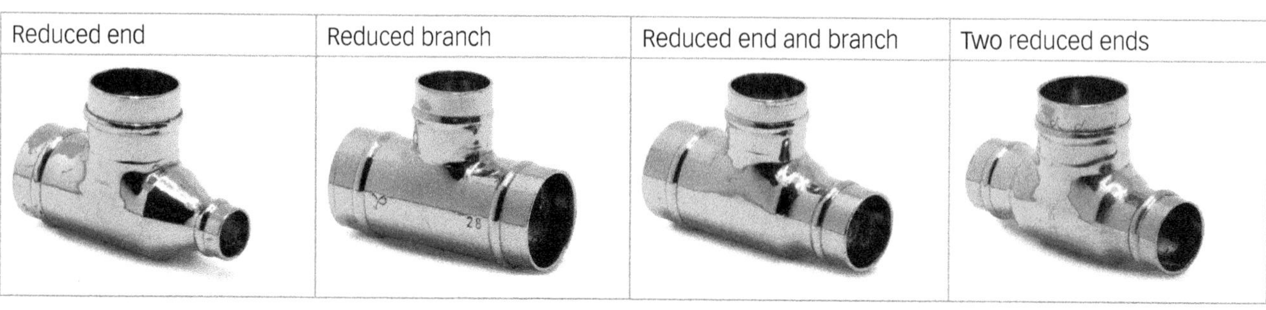			

Table 1.26 Cap ends

Compression cap end	Push-fit cap end	End feed capillary cap end

Table 1.27 Connectors and manifolds

Tank connector	Flexible connector	Manifold
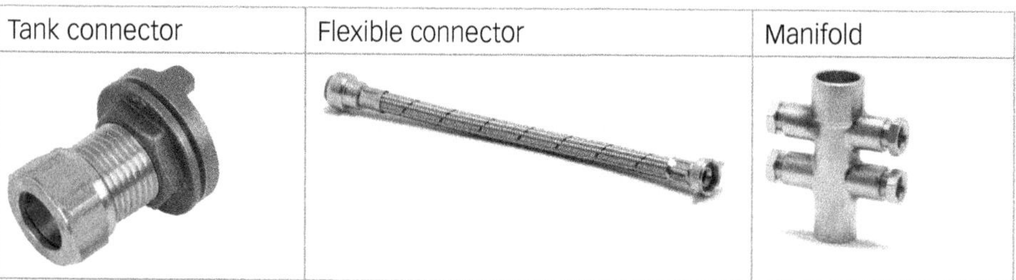		

Table 1.28 Tap connectors

Straight tap connector	Bent tap connector

Exam tip

You might be asked in an exam about re-using fittings. The end feed, solder ring and press-fit fittings are one use only. The compression and push fit fittings can be used again.

Table 1.29 Copper fittings

End feed		+ These fittings require cleaning along with the copper tube + The flux wiped on to the tube + Heated up by a blow torch + Solder fed into the fitting
Solder ring		+ These fittings require cleaning along with the copper tube + The flux wiped on to the tube + Heated up by a blow torch + Solder is already in the fitting – note the ring containing the solder

Leaded solder	+ Only used on central heating installation	
Lead free solder	+ Used on any installation	
Active flux	+ Known as 'self-cleaning' flux because of the acidic content + Must be cleaned off and flushed out after soldering	
Traditional flux	+ Cleaning of the fittings and pipe MUST be carried out properly	
Compression (type A)		+ Uses a nut and olive to create the water-tight seal + This is a standard compression fitting + Commonly used above ground
Compression (type B)		+ The end of the copper pipe needs to be 'swagged' out and an adaptor inserted + This is a special compression fitting used for high pressure installations (underground water mains or steam) + Commonly used on below ground mains

Check your understanding

9 Why must leaded solder not be used on hot and cold water installations?

10 Where is it stated that leaded solder must not be used on potable water systems?

Table 1.30 Proprietary copper fittings

Leadlok		These are specially made to connect lead pipe to copper pipework
Philmac		These are used to connect the incoming MDPE cold water main pipe to the internal copper pipework

Proprietary fitting
A fitting 'made for the purpose of' something.

Typical mistakes

When asked to identify a fitting, do not make any assumptions. Make sure that you look at the image carefully and identify it correctly.

Table 1.31 Plastic pipes

PVCu	+ Unplasticised polyvinyl chloride + Used for push fit and solvent welded pipework and fittings + Soil and waste pipe; guttering and downpipe + 110 mm, 50 mm, 40 mm, 32 mm
ABS	+ Acrylonitrile butadiene styrene + Solvent welded pipework and fittings + Soil and waste pipe + 110 mm, 50 mm, 40 mm, 32 mm
MUPVC	+ Modified unplasticised polyvinyl chloride + Solvent welded pipework and overflow pipes + More durable than PVCu + Degrades in daylight/sunlight + 110 mm, 50 mm, 40 mm, 32 mm, 19 mm

Polypropylene	+ Used for waste pipe + Push fit only + 50 mm, 40 mm, 32 mm
Polyethylene (MDPE)	+ Medium density polyethylene + Blue in colour + Degrades in sunlight + Underground cold water mains pipework
Polybutylene	+ Plastic pressure pipe bought in long coils + Push fit or compression fittings + Domestic central heating, hot water and cold water systems

Plastic pipe fittings

Table 1.32 Common ring seal soil pipe fittings

90° bends		Strap boss	
45° bends		Access pipes	
Junctions		Boss pipe adapters	
Sockets		Pipe clips	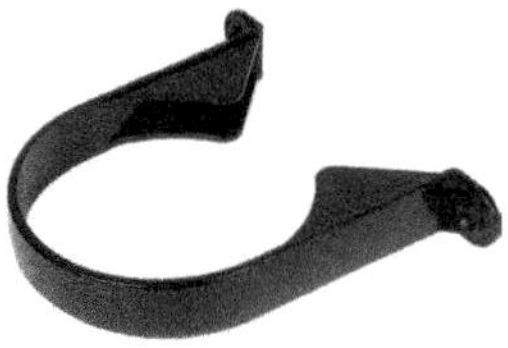
Boss pipes		Waste socket	

Table 1.33 Common solvent weld fittings

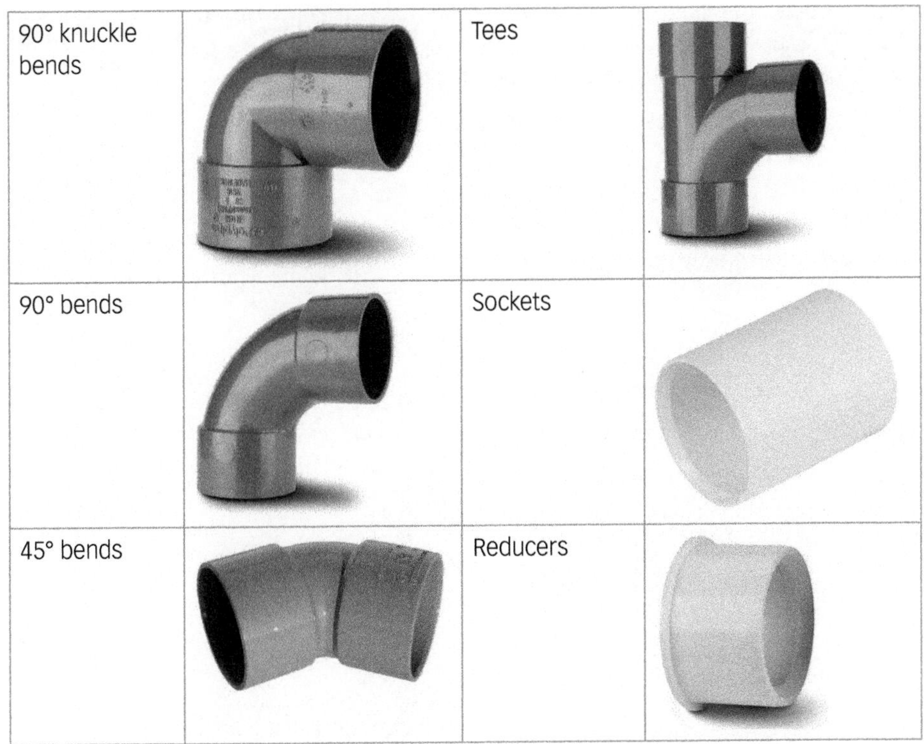

90° knuckle bends		Tees	
90° bends		Sockets	
45° bends		Reducers	

Table 1.34 Common ring seal waste pipe push-fit fittings

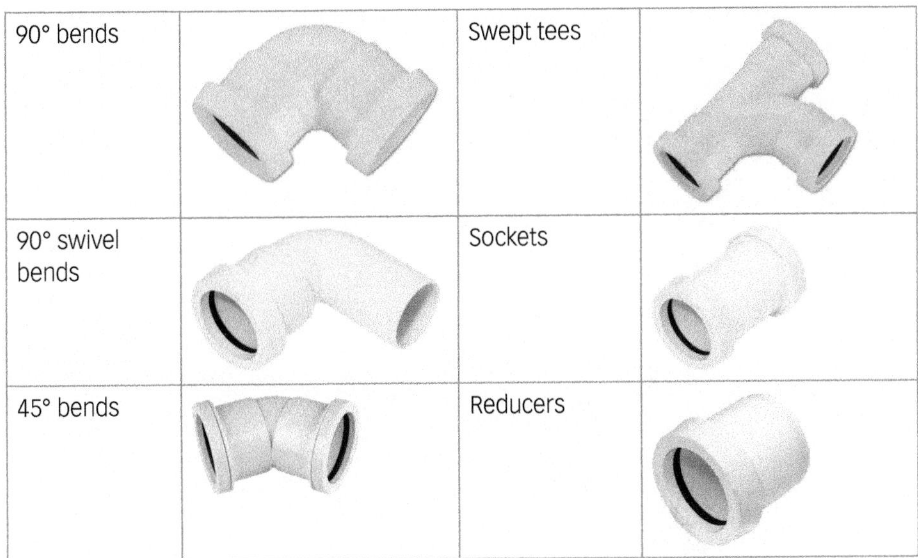

90° bends		Swept tees	
90° swivel bends		Sockets	
45° bends		Reducers	

Table 1.35 Plastic pressure pipe push fit fitting

Hep2O	Speedfit	Polyplumb

When installing plastic pressure pipe (polybutylene), an insert **must** always be used to offer support to the pipe, whether push fit fittings or compression fittings are being used.

These are pushed into the end of the pipe prior to the fitting being installed.

The push fits for PPP also have grab rings to retain the pipe in position.

Table 1.36 Low carbon steel pipe (LCS)

Light grade		Gas pipework
Medium grade		Most common general pipework
Heavy grade		High pressure and steam pipework

Exam tip

Questions may come up about which LCS grade is used for certain installations. Blue grade (medium) is the LCS pipe that is used for central heating pipework.

Check your understanding

11 What **must** an LCS pipe be protected by if it is installed in a cold water system?

Table 1.37 LCS pipe threads

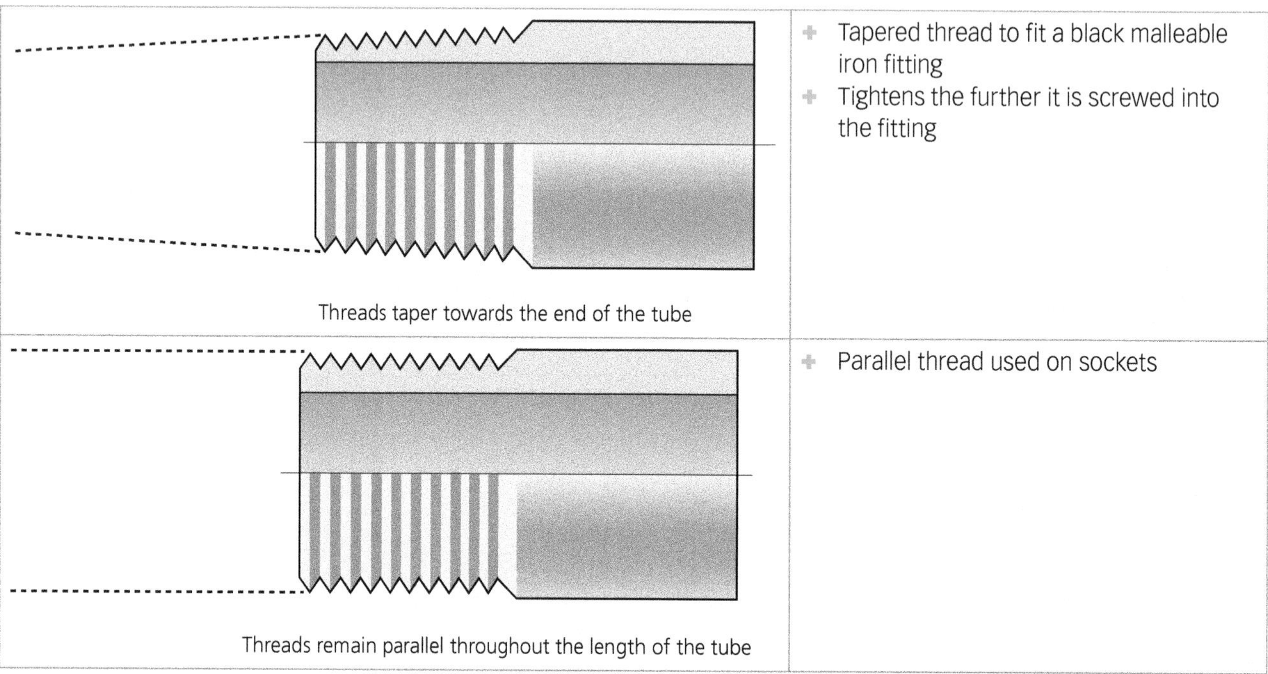

Thread	Description
Threads taper towards the end of the tube	+ Tapered thread to fit a black malleable iron fitting + Tightens the further it is screwed into the fitting
Threads remain parallel throughout the length of the tube	+ Parallel thread used on sockets

LCS pipe fittings

Table 1.38 Common LCS fittings

Couplings		Unions	
Equal tees		Nipples	
Elbows		Bushes	
M/F elbows			

Table 1.39 Clipping distances

		Max horizontal	Max vertical
Copper	15 mm	1.2 m	1.8 m
	22 mm	1.8 m	2.4 m
Plastic (Polybutylene)	15 mm	0.3 m	0.5 m
	22 mm	0.5 m	0.8 m
Low Carbon Steel (LCS)	½"	1.8 m	2.4 m
	¾"	2.4 m	3.0 m

Table 1.40 Watertight joints in all fittings

Copper	+ Solder – quality of the soldering + Push fit – O ring + Compression – the olive
Plastic	+ Push fit – O ring + Compression – O ring + Solvent weld – quality of welding
LCS	+ Thread – hemp and paste or **PTFE tape** + Union – surface finish or gasket + Compression – O ring

Now test yourself TESTED

6 You are installing a long length of cold water pipework in a property and would like as few connections as possible to avoid potential leaks. What pipe would you install?

7 You are pricing up a new soil stack at a customer's house and you would like to use a mixture of solvent welded and push fit fittings for the main stack. However, you need to keep to one type of plastic. What grade of plastic could be used?

Exam tip

In an exam, questions may ask about specific clipping distances **and** why these distances are used. Note how much closer the polybutylene clipping distance is. This is due to the plastic pipe becoming a lot more flexible when hot water passes through, so it is prone to sagging.

Typical mistakes

Not being able to recall the correct clipping distances. Vertical is upright and horizontal is level (think of the horizon). The copper and LCS clipping distances are related to the 6 times table – 12 /18; 18/ 24; 24 /30.

PTFE tape Polytetrafluoroethylene tape (or plumber's tape for everything).

Topic 4.3 Bend pipework work for installation

REVISED

This is a workshop activity in which you will need to show that you know how to bend pipework work for installation, including the use of the following:

+ types of copper machine bending (90 degree bends, sets and offset bends, Passover bends)
+ types of LCS hydraulic machine bending (90 degree bends, sets and offset bends, Passover bends)
+ cabling technique for plastic pipework.

Topic 4.5 Install pipework

REVISED

This is a workshop activity in which you will need to show that you know how to install pipework, including the use of the following:

+ how to install different types of pipework (select, measure, mark-out, cut, joint, bend, fabricate, fix and test pipework)
+ types of pipework to be installed (copper, LCS and plastic).

Exam-style questions

1 When installing an outside tap at a customer's property, the customer was asking why the pipe was sleeved. What should you say?
   a To allow for movement
   b To stop air flow
   c To avoid moisture
   d To prevent corrosion

2 What class of copper pipe would be used if you were installing a new central heating system in a new build?
   a R300 special copper
   b R220 soft copper
   c R250 half hard copper
   d R290 hard copper

3 What creates the water-tight seal in a PUSH FIT fitting?
   a Olive
   b Hemp and paste
   c Neoprene O ring
   d Brass ring

4 Your supervisor hands you a circular saw, but asks you what you need to check before you use it. Which one of the following would you say?
   a MCB rating
   b Guard is present and working
   c The grip handle is non-slip
   d The push plate is secure

5 Which part of the scissor type bender is the GUIDE or SLIP?

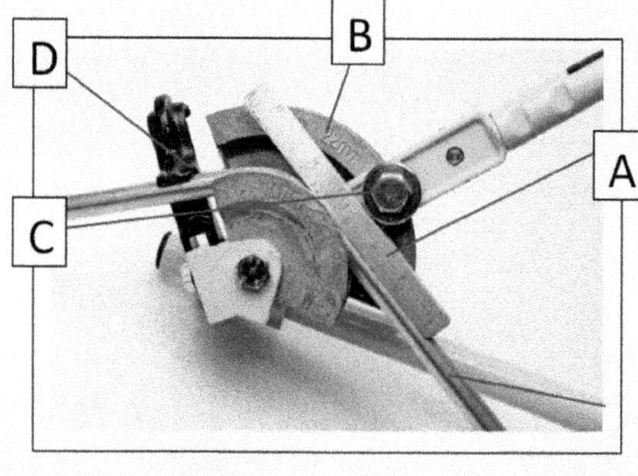

   a A   b B   c C   d D

6 What material would a self-drive cavity fixing be used in?
   a Wood   c Concrete
   b Masonry   d Plasterboard

7 You have to chase along a wall to hide some pipework. Your supervisor asks why you have to be careful about the depth of a horizontal chase. What is the reason?
   a The amount of dust created when chasing
   b The vibration when chasing can crack the plaster
   c The depth could make the wall unstable
   d A more powerful tool will be required

8 What creates the water-tight seal on a threaded Low Carbon Steel connection?
   a Rubber O ring   c Hemp and paste
   b Olive   d Brass seal

9 What material is this drill bit used in?

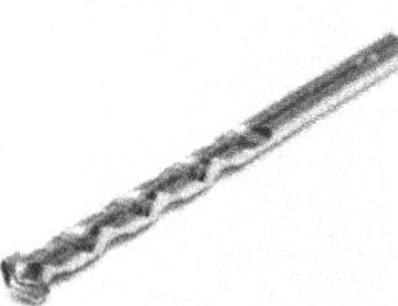

   a Masonry   c Plasterboard
   b Wood   d Metal

10 You are using an adjustable spanner on a compression fitting. As you tighten the fitting up the spanner slips off the fitting. What should you do?
   a Replace the adjustable spanner
   b Attend a training course
   c Lubricate the worm and wheel
   d Adjust the jaws correctly

11 What material can a pipe wrench with a footprint-type grip be used on?
   a Brass   c Plastic
   b Copper   d LCS

12 Why are countersunk screws used when securing floorboards back down?
   a They do not corrode
   b They are aesthetically pleasing
   c The head goes flush with the floorboard
   d The head does not mark the carpet when re-laid

13 What is the maximum depth a notch can be cut to in a joist?
   a 1/4 of the joist depth
   b 1/6 of the joist depth
   c 1/8 of the joist depth
   d 1/10 of the joist depth

14 When must you not use an SDS drill to drill a hole in a wall?
   a When it has a green PAT test label
   b When it is brand new
   c When the lead is frayed
   d Before your supervisor has tested it

15 What is the correct way to replace a hacksaw blade?
   a Teeth facing away with no tension
   b Teeth facing you under tension
   c Teeth facing you with no tension
   d Teeth facing away under tension

16 Where might a plumber use a mirror screw?
   a Floorboard fixing
   b External fixing
   c Bath panel
   d Securing plasterboard

17 When drilling a hole in a joist so a pipe can be laid across a room, where in the joist must that hole be placed?

- **a** Top 25%
- **b** Bottom 25%
- **c** Not within the top or bottom 20 mm
- **d** Centre line

18 What pipework can an interlocking clip secure?

- **a** Copper, plastic and LCS pipework
- **b** Copper and plastic
- **c** Copper only
- **d** LCS and plastic

19 What power tool would you use to make a hole in the wall for a new flue when using a core drill?

- **a** Rotary hammer action drill with hammer action
- **b** SDS hammer drill with no hammer action
- **c** Battery drill
- **d** Self-centring chuck drill with hammer action

20 What item prevents smoke and fire spreading between floors in a building?

- **a** Intumescent collar
- **b** Compression seal
- **c** LCS union
- **d** Copper pipe reducer

21 Which of the following fittings requires solder to be fed in to create the watertight seal?

- **a** Solder ring
- **b** Push fit
- **c** End feed
- **d** Press fit

22 What is the main hazard when using a wood chisel to notch out a joist?

- **a** The handle splitting
- **b** Mushroom head
- **c** Very sharp cutting edge
- **d** Hitting your finger

23 You are installing a commercial central heating system in a school. What grade of LCS would be installed?

- **a** Brown
- **b** Red
- **c** Green
- **d** Blue

24 What does the works or job schedule do?

- **a** Monitors progress on site
- **b** Lists the materials used on a specific installation
- **c** Outlines the number of staff required
- **d** States where to install appliances

25 When installing a new water main to a property, you have a choice of plastic pipework. Which is the correct plastic to install?

- **a** Polybutylene
- **b** ABS
- **c** MDPE
- **d** PVCu

2 Electrical and scientific principles (Unit 213)

This unit covers properties of materials, liquids and gases; heat transfer; pressure and forces along with basic concepts of electricity. It looks at their impact on plumbing systems. Try not to focus on the detail but concentrate on your weaker areas in this chapter. Ask yourself:

- What is heat and how is it used?
- How do I measure pressure and where is it in a system?
- What materials are suitable for a plumbing system?

Typical mistakes

Thinking of this subject as pure science; relating each element to a plumbing situation will help you to recall the information.

LO1 Understand materials used in the plumbing industry

Topic 1.1 Material properties used in the plumbing industry

REVISED

Materials that you will use in the plumbing industry have various physical properties, including:

- **tensile strength** – ability of a material to be pulled or stretched

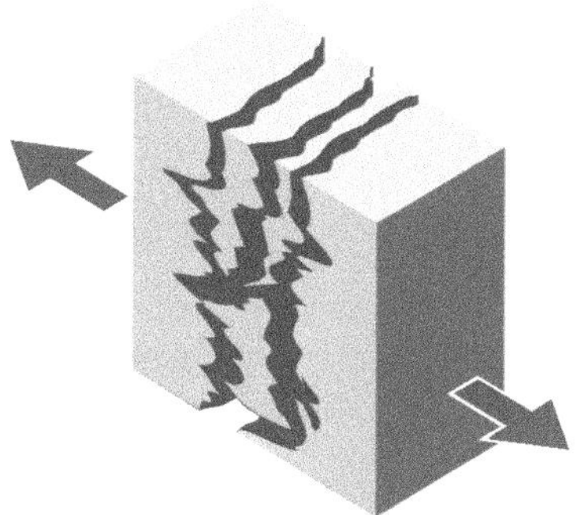
Tensional stress

Exam tip

To remember tensile strength, think of a munsen ring hanging from the ceiling holding up some pipework.

- **compressive strength** – ability of a material to withstand being crushed

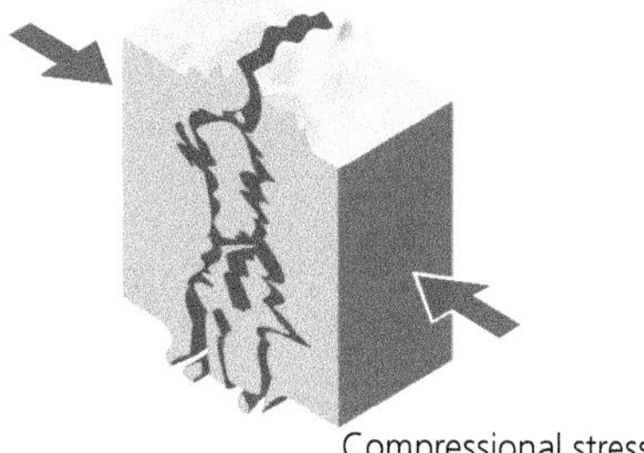
Compressional stress

Exam tip

To remember compressive strength, think of a leg under a bath and the pressure the legs are under when the bath is full of water.

- **hardness** – ability to resist scratching, denting or cutting. Diamond is the hardest material
- **ductility** – ability to be pulled, pushed and stretched into shape

Exam tip

To remember ductility, think of bending copper pipe – copper is very ductile.

- **malleability** – ability to be formed by compressive force (for example, hit, hammered or bossed)
- **conductivity**:
  - thermal – where heat is transferred. Good heat transfer, good thermal conductivity
  - electrical – where an electrical current can be transferred. Good transfer, good electrical conductivity. Copper has very good heat and electrical conductivity (for example, copper hot water cylinder and coil; copper wires).

Exam tip

Lead is very malleable and can be bossed into shape without breaking.

Classification of metals

Ferrous – they contain iron and so are magnetic (for example, cast iron and low carbon steel pipe). These metals can rust and will need protection.

Non-ferrous – they do not contain iron and so are not magnetic (for example, copper, lead, zinc, aluminium). These will not rust but can corrode over time.

Alloys – mixture of two or more metals, such as:

- brass – copper and zinc
- bronze – copper and tin
- low carbon steel – iron and carbon
- lead free solder – tin and copper.

Check your understanding

1 Which common plumbing material is used because of its malleability and good thermal conductivity?

Thermal expansion

This is how much a piece of material expands by when heated. The heat could be caused by many things including the sun, a flame, hot water passing through and so on.

Plastics

- These expand the most out of all materials.
- The coefficient of linear expansion is 0.00018.

Common materials

- The coefficient of linear expansion of common materials includes:
  - lead – 0.000029
  - copper – 0.000016
  - steel – 0.000011.

(Note there are four zeros after the decimal point!)

Pipes

- When pipes are heated, they expand.
- When pipes cool down, they contract.
- This movement can create a creaking noise under the floor if notching is too tight. Expansion joints may have to be included in larger installations.

Guttering

- To allow for expansion, manufacturers build in a 10 mm expansion allowance. The gutter is inserted up to the expansion mark, not to the stop.

Coefficient of linear expansion All materials expand by a small amount in length when heated. The amount they expand by is measured in millimetres. This number is known as the coefficient of linear expansion: it is a measure of how much a material expands by for every degree C it heats up, per metre of pipe or material used. The heat could be caused by many things including the sun, a flame, or hot water passing through.

You can calculate this by:

length of pipe × coefficient of expansion × temperature rise.

So, for 5.0 m of plastic guttering in summertime when the temperature rises 150C: 5000 mm × 0.00018 × 15 = 13.5 mm expansion.

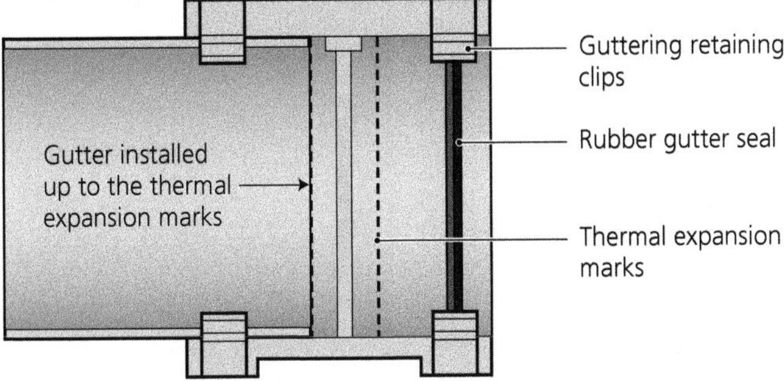

Typical mistake

For linear expansion, make sure you are aware of how many zeros there are after the decimal point. With plastics there are three zeros – 0.00018 (greater expansion) but with metals like copper, there are four zeros – 0.000016 (less expansion).

Check your understanding

2 What is the major problem with ferrous material that requires our attention?
3 Which material expands the most: steel, plastic or copper?

Exam tip

When you work out the expansion using the equation, it is the temperature rise that is used, **not** the start or end temperature. For example, the weather starts at 15°C in the morning, rising to 25°c in the afternoon. By how much will the 5.0 m (convert to 5000 mm) length of plastic gutter expand? Your answer must use the temperature rise of 25 – 15 = 10°C.

Topic 1.2 Uses of materials

REVISED

There are different types of materials that you will come across in industry and they all have different uses.

Types of metals and their uses

Table 2.1 Types of metals and their uses

Steel	Radiators are made from pressed steel. Baths and boiler cases can also be made from pressed steel
Iron	Central heating circulators are made from cast iron. Munsen rings and older or decorative column radiators, old soil stacks and guttering can also be made from cast iron.
Copper	Domestic pipework and hot water cylinders are made from copper
Lead	Old pipework may still be made of lead It is not allowed to install any lead pipework or use leaded solder on hot and cold systems
Stainless steel	Stainless steel screws used to secure fascia brackets as they will not corrode Heat exchangers can be made from stainless steel
Brass	Compression fittings and valves are generally made from brass **DZR** brass important as this is de-zincification resistant
Solder	Leaded is only allowed on central heating systems, so is not favoured as it can contaminate water systems. Lead-free solder is the choice to be used.

DZR The letters used on new brass fittings to identify de-zincification resistance.

Types of plastics and their uses

Thermoplastics – can be recycled. They are brittle when cold. Most common plastic plumbing components.

Thermosetting – once set, these plastics cannot be remoulded (for example, light switches).

uPVC – (chlorinated) unplasticised polyvinyl chloride.
- Solvent weld and pushfit soil pipes and fittings.
- Underground drainage pipework and fittings.
- Guttering and downpipes.

MDPE – medium density polyethylene.
- Blue underground water mains pipework.

Polypropylene
- Push fit waste pipework.
- Cold water storage cisterns.

Polybutylene
- Push fit pipework for hot, cold and central heating systems.
- Known as plastic pressure pipe (PPP).

ABS – acrylonitrile butadiene styrene
- Solvent weld pipework.

Check your understanding

4 Which plastic is used on central heating systems because it can withstand higher temperatures?

Types of ceramics and their uses

- **Fireclay** – heavy duty appliances, like cleaners' sinks (Belfast and London sinks).
- **Vitreous china** – enamelled coated and used for bathroom appliances, for example, WC pans, basins, bidet.

Topic 1.3 Corrosion protection and degradation

REVISED 

The materials that you use in installations will be slowly damaged over time by elements such as water, heat, oxygen and UV light. This can corrode or degrade the material making it unfit for use, so you must be aware of the possible problems outlined in the table.

Table 2.2 Corrosion and degradation

Atmospheric conditions	+ All material corrodes over time + The humidity (water) in the air and the oxygen in the air are key factors
Oxidisation	+ Metals oxidise due to the presence of oxygen + More commonly known as 'corrosion' + Oxygen penetrates the metal surface, causing the growth of oxidation
Rusting	Iron + oxygen + water = rust
Dezincification	+ Zinc is leached out of brass as oxygenated water passes through + White powder left, with component being very brittle + DZR brass fittings are de-zincification resistant
Electrolytic corrosion and principles	+ Also known as 'galvanic corrosion'. It happens when dissimilar metals are connected together and the system is filled with water (electrolyte) + Copper pipe, brass fittings, stainless boiler, cast iron pump could easily be connected together in the same system + The system water becomes the electrolyte (the system becomes a battery) + Weakest metal is eaten away – a hole forms + Copper is the strongest metal, magnesium is the weakest metal + Cathode eats the anode + This is why the 'sacrificial anode' in a hot water cylinder is made of magnesium
Erosion	+ Occurs in pipework and fittings due to the fast flow of liquids + Increased turbulence caused by burrs can cause erosion + Pipework and fittings are worn away
Pitting	+ Type 1 pitting is caused by a system not being flushed out correctly + Flux is left inside the pipework, which eats away causing a pin hole leak + This can also be known as 'pinholing'
Degrading	This happens to plastics exposed to the atmosphere: + thermal (e.g. heat from the daytime temperature) and waste water + light from the sun + oxygen ingress from the atmosphere + ultraviolet (UV) from solar radiation

There are also methods you can use to prevent corrosion, such as:

Galvanising

- Is used on LCS pipework that carries water (cold water mains).
- This is a zinc plating that coats both inside and outside the pipe.

Coatings

- Copper pipe is sometimes coated in a plastic to protect from corrosion.
- LCS pipework and cast-iron guttering need to be painted on the outside to protect from rust.
- Copper pipe is sometimes chrome-plated for aesthetic reasons.
- Steel screws are anodised to prevent corrosion.
- LCS pipework can be covered with an oil-based bandage to protect it from corrosion.

Corrosion inhibitors

- Chemicals are added to a central heating system to prevent magnetite.
- The inhibitor level should be tested regularly so the correct levels are maintained.

Magnetite A form of rusting or oxidation on the inside of the central heating system. Also known as 'black sludge'.

Sacrificial anodes

- Found in a hot water cylinder or boiler.
- Made of magnesium (weak metal).
- Prevent electrolytic corrosion from attacking the system metals.

Now test yourself TESTED

1 You are called to a customer's property where a combination boiler is continually losing pressure. On inspection, you find a damp area below some copper pipe near to a soldered connection. The connection appears fine, but the weep is coming from the pipe where there is green build up from the flux. What type of corrosion is likely to have caused this leak?

LO2 Understand properties of water, liquids and gases

Topic 2.1 Properties of water

REVISED

Water has the following properties:

- Colourless.
- It changes state – **liquid** (water), **solid** (ice), **gas** (steam).
- At its **most dense at 4°C.**
- **Boiling point 100°C** (at sea level), under pressure the boiling point increases.
- **Freezes at 0°C**, anti-freeze (glycol) can be added to prevent freezing.
- When frozen, it **expands by 10%** (hence burst pipes.)
- **Relative** density of 1.
- Specific heat capacity of **4.187 kJ/kgK** (takes a lot of energy to heat up).
- Can be **hard (alkaline)** or **soft (acidic)** with different pH values.
- When heated in a system (between 10 and 99°C) it **expands by 4%.**
- If allowed to turn to **steam**, it **expands by 1600 times.**
- **Capillarity** can take place between two close fitting surfaces, which means water can move up or through the close fitting surfaces.
- **Super-heated steam** can be produced and used in certain commercial systems like steam turbines.

Exam tip

You may be asked a question on the changes of state of water. Remember these are due to molecular changes.

- **Ice** – the molecules are close together.
- **4°C** – the molecules are as close as they will ever get (maximum density).
- **Liquid** – the molecules are freely moving.
- **Steam** – the molecules are far apart.

Check your understanding

5 When a hot water cylinder is heated up, the water will expand by approximately 4%. Where is that expansion taken up?

Exam tip

In an exam you may be asked to work out the expansion volume of water when it is heated in a hot water cylinder. Be prepared to work out 4% of a volume of water in a hot water cylinder (for example, 4% of 100 or 125 litres).

Topic 2.2 Properties of liquids

REVISED

Different liquids have the following properties:

Table 2.3 Properties of liquids

Water	Capillary action: + Water can be drawn upwards between two close fitting surfaces (against gravity) + The wider the gap, the less capillary action + This also happen when soldering a copper fitting. The solder is drawn between the fitting and pipe + Can be a negative with an **S trap**, if the trap seal is drawn out Adhesion and cohesion: + Adhesion happens when water wants to stick to a surface like the edge of a glass – curved edge + Cohesion happens when the water molecules stick to each other and causes surface tension (e.g. when you overfill a glass)
Refrigerant	+ Fluorinated chemicals used in liquid and gas states + Compressed – it is a liquid + Pressure-released – it is a gas + Boils well below freezing point + Used in air conditioning and heat pumps
Anti-freeze (Glycol)	+ Used to prevent water from freezing at 0°C + Added to solar thermal systems that have roof line pipework; to heat pumps with pipework deep in the ground
Fuel oil	+ Also known as 'kerosene' + It is a thin, clear hydrocarbon oil + Used for domestic oil boilers
Lubricants and greases	+ Used between two moving parts to maintain (or ease) movement + Silicon grease used to lubricate rubber seal on fittings + Graphite paste used on gas taps + Cutting oil used when threading LCS pipework + Penetrating oils used to loosen items

S trap A style of trap used under a sanitary appliance and derives its name from its shape.

Exam tip

Questions on refrigerants could be asked, so it is important to remember the basics about these liquids. If compressed, they are a liquid. Otherwise, they are a gas and refrigerants boil well below the freezing point of water.

Topic 2.3 Properties of gases

REVISED

Different gases have the following properties:

Table 2.4 Properties of gases

Air	+ Limited use within plumbing + Warm air heaters use air to transmit heat + Used to recharge the dry side of an expansion vessel + High pressure air is used to clear blockages
Steam	+ Used to be used a lot within commercial heating + Combined heat and power systems use steam to produce electricity + Power stations use steam to produce electricity + Hot water is produced in calorifiers
LPG	+ Liquid petroleum gas + Used in heating appliances like boilers, cookers and fires (butane) + Used for soldering (propane) + Delivered in bottles

Natural gas	+ Widely used domestic fuel + Used for heating appliances like boilers, cookers + Delivered from the national grid
Carbon dioxide	+ Used as freezing agent when pipe freezing + Used in the fire extinguishers used by plumbers

Typical mistakes

Questions will be asked in the exam about properties of liquids and gases, so do spend time revising this area.

LO3 Understand density, force, pressure, flow rate and basic mechanics

Topic 3.1 Types of SI units

REVISED

As a plumber you will have to use various units of measurement when installing, marking out, interpreting a drawing, reading manufacturer's instructions and commissioning systems. This means you will need to have a good knowledge of what these units are associated with. SI units are used as a standard across industries and countries.

Table 2.5 Units of measurement

Length	Metre (m)
Mass	Kilogram (kg)
Time	Second (s)
Temperature	Kelvin (K)
Force	Newton (N)
Pressure	Pascal (Pa)

Table 2.6 Applications and use of SI units

Area (length × width)	m^2
Volume (length × width × height)	m^3
Litres (1000 litres = 1 m^3 @ 4°C. Litres can also be used to measure flow rate in l/s.)	l
Density (mass over volume)	kg/m^3
Velocity (distance in a second)	m/s

Check your understanding

6 After installing a new shower, the manufacturer's instructions state that you must test the flow rate through the shower using a weir cup. What unit of measurement would be used?

Topic 3.2 Density of materials

REVISED

Density refers to the relationship between the mass of a substance and its size. Materials will have different densities depending on their properties, for example, 1 litre of water = 1 kg. They will also have a **relative density**, which is the mass of a particular volume of a substance when compared with the mass of an equal volume of water at **4°C** or air.

Relative density:

+ Water and air have a relative density of one (1).
+ Anything with a relative density **higher than 1** will sink or fall.
+ Anything with a relative density **lower than 1** will float or rise.

Table 2.7 Relative density of materials

Relative density of water = 1	
Copper	9
Steel	7.5
Lead	11.3
Brass	8.4
uPVC	1.35
Polypropylene	0.91

Now test yourself TESTED

2 Out of the materials listed in Table 2.7, which one would a) float and b) sink the quickest?

Table 2.8 Relative density of gases

Relative density of air = 1	
Natural gas	0.6
Propane	1.5
Butane	2.0

Now test yourself TESTED

3 Out of the gases listed in Table 2.8, can you name any that would cause a hazard if used in a cellar?

Topic 3.3 Force, pressure and flow rate

REVISED

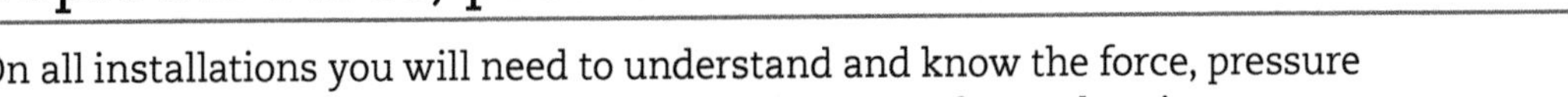

On all installations you will need to understand and know the force, pressure and flow rate of the system. These checks will have to be made prior to an installation and at the commissioning stage.

- Force on an object results in movement (for example, gravity acting on water stored in a cistern).
- Pressure in a system can be increased by adding a pump or increasing the head height of the cistern.
- Flow rate is the amount of water coming out of an appliance. It can be increased by using a larger pipe diameter.

Table 2.9 Units of force and pressure

Acceleration	How much something's velocity (m/s) increases by in 1 second (m/s/s or m/s^2)
Force	Newtons (N)
Pressure	Force over an area (N/m^2)
Flow rate	Litres per second (l/s)
Gravity	9.81 N/m^2
Atmospheric pressure	101.3 N/m^2
Bar	+ Used when pressurising an expansion vessel + Used when measuring pressure of water at an outlet + When measuring natural gas, millibars (mBar) are used as natural gas is low pressure
kPa (Kilo pascals)	+ Scientific way of measuring pressure at an outlet + Used in manufacturer's instructions

Exam tip

If you are asked how much water comes out of a tap in a minute, all you do is multiply the litres per second by 60 (60 seconds in a minute).

PSI	+ Pounds per square inch + Imperial way to pressurise an expansion vessel
Metre head When the tap is closed, the body of water is at rest	+ Distance between the water level in the cistern and the outlet point (as the water level varies the base of the cistern is sometimes used as a constant) + The greater the head height, the greater the outlet pressure
Flow rate	+ How much water comes out of the outlet in a certain time – litres per second (l/s) or litres per minute (l/min) + Measured with a flow cup

Typical mistakes

In an exam question, it might refer to a flow cup or a weir cup. Don't forget they are the same thing!

Check your understanding

7 What is the force in Newtons due to gravity?

Typical mistakes

Remember, if you are asked to define pressure: pressure is FORCE (N) over AREA (m^2).

Table 2.10 Pressure head conversion table: pressure is force over area (N/m^2), which is why Pascal is a measurement of pressure ($1 Pa = 1 N/m^2$)

Kilopascals (kPa)	Bars	Metres head of water	Pounds per square inch (psi)
10	0.1	1	1.42
20	0.2	2	2.84
30	0.3	3	4.27
40	0.4	4	5.68
50	0.5	5	7.11
100	1	10	14.22
150	1.5	15	21.33

Exam tip

Remember that 1 litre of water weighs 1 kg. So, if 8 litres of water flows from an outlet in one minute, you have 8 kgs of water each minute!

+ Note as the head height increases, so does the pressure.
+ Force is an influence on an object that will cause movement.
+ Gravity is pushing down at 9.18 N/m^2.

Check your understanding

8 You check the pressure at a gravity shower outlet and the gauge reads 0.6 bar. How high would the storage cistern need to be to offer this pressure?

Now test yourself TESTED

4 A customer complains of low water pressure at the upstairs basin. What two options could be offered to the customer to improve the pressure and flow rate at this outlet?

- So, the weight of the water along with gravity makes the water in the system move downwards.

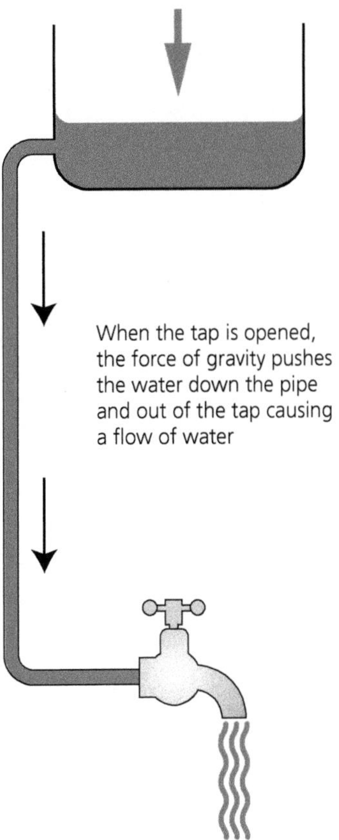

Force of gravity on water

Now test yourself TESTED

5 Which of the following cold water cisterns will apply the most pressure onto the loft floor, if the force from each cistern remains the same because the volume is the same? (Remember: Force / area = N/m^2.)

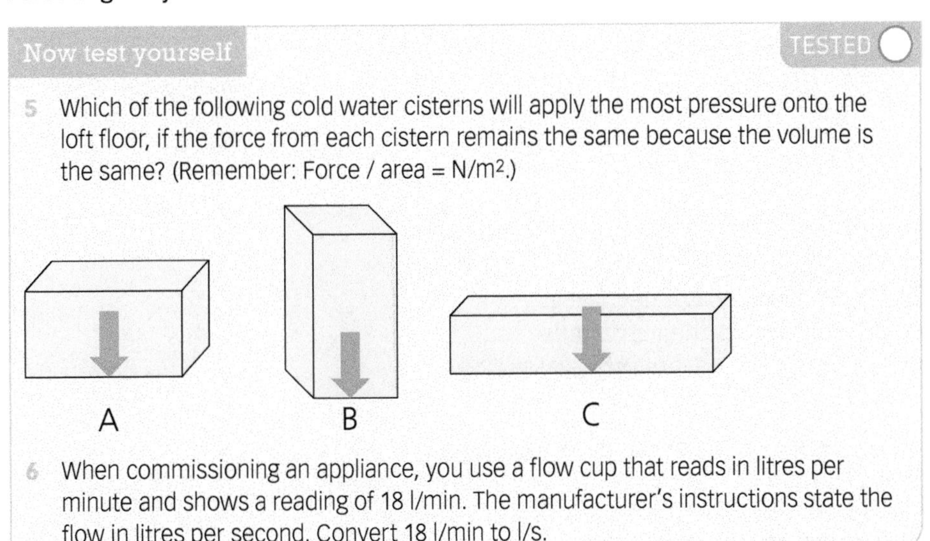

6 When commissioning an appliance, you use a flow cup that reads in litres per minute and shows a reading of 18 l/min. The manufacturer's instructions state the flow in litres per second. Convert 18 l/min to l/s.

Increasing the pressure and flow:

- Increase velocity.
- Increase flow rate.
- Pressure can be increased by adding a pump to the system or increasing the head height.
- Flow rate can be increased without increasing the pressure – increase the pipe size.

Decreasing the pressure:

- Decrease the velocity.
- Decrease the flow rate.

Factors affecting flow rates:

- Changes in direction – elbows offer greater resistance than machine bends. Swept tees should be used to aid flow.
- Pipe size – smaller pipes less flow; larger pipes more flow.
- Pressure – the higher the pressure, the higher the flow.
- Length of pipe – in a longer pipe, the flow rate will diminish.
- Frictional resistance – the smoother the pipe (plastic) the better the flow; the rougher the pipe (galvanised LCS) the worse the flow.
- Constrictions – (like valves) offer great restrictions to the flow of water.

Typical mistakes

Not being able to apply and relate these factors to a question in the test, so make sure you have plumbing scenarios in mind when these factors come up. For example: installing too many elbows; 15 mm copper low pressure supplying a bath; head height from the cistern; the choice of materials and the correct valves.

Check your understanding

9 Which isolation valve can only be installed on a low pressure system?

Now test yourself TESTED

7 You have a choice to make on site: one installation has three elbows and two machine bends, the second has five machine bends. Which installation would offer the better flow of water?

Topic 3.4 Mechanical principles

REVISED

Mechanical principles allow the efficient execution of movement in many situations. These principles are used when choosing the correct tool to use, keeping your balance up a ladder, lifting or moving heavy weights.

- **Simple machines** include levers and pulleys.
- **Basic mechanics** include movement of force, equilibrium and centre of gravity.
- **Theory of moments** – what makes an object move, pulling, pushing, lifting, lowering using a force.

Action and reaction:

- A push or pull on an object (action) can make the object move (reaction) – **contact force**.
- If the action and reaction are equal, there will be **no** movement because the forces are **equal**.
- Contact forces – normal, friction, tension and applied forces.
- Other forces – gravitational pull, electrical or magnetic pull.
- **Newton's third law** – every action has an equal but opposite action.

Centre of gravity:

- An imaginary point where all the weight of an object is concentrated.
- It will vary from object to object.
  - Symmetric objects – the centre of gravity will be the geometric centre.
  - Irregular-shaped objects – it could be outside the object.
  - Hollow objects – (such as a football) it is in free space.

Equilibrium:

- All forces acting on a stationary object are balanced – 'state of equilibrium'.
- The forces are balanced when (left, right; front, back; up, down) are the same.

Equilibrium Balanced

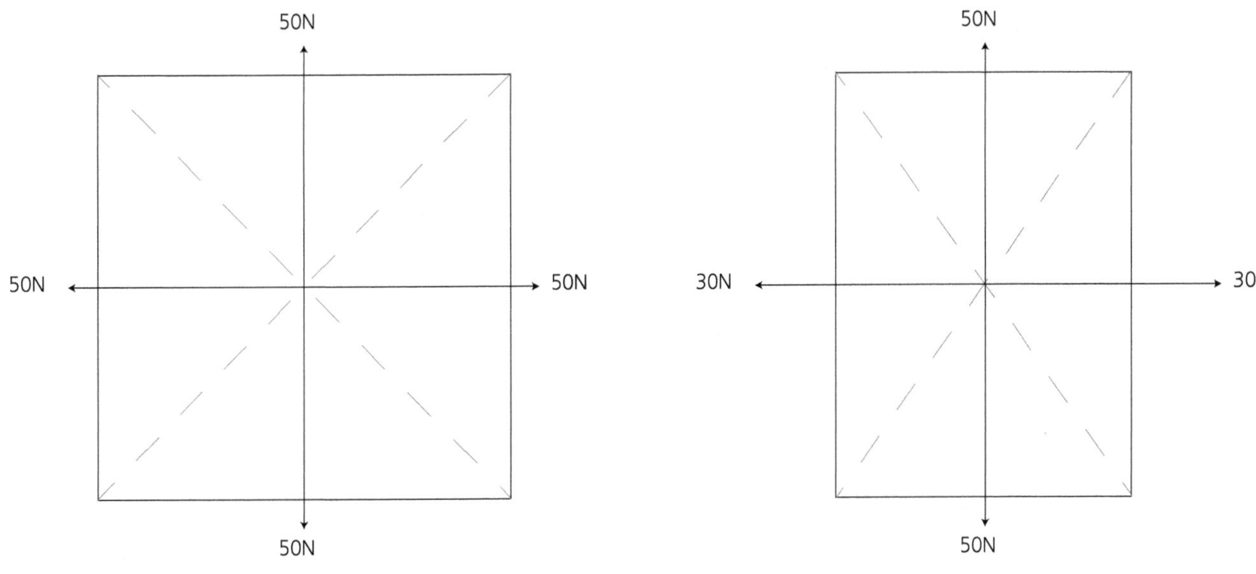

Balanced and unbalanced forces in equilibrium

Velocity ratio:

$$\frac{\text{distance moved by lifting force}}{\text{distance moved by load}}$$

- Think of a crowbar, where the distance moved by the handle is far greater than the distance moved by the load. This shows the ratio of force to load.

Mechanical advantage:

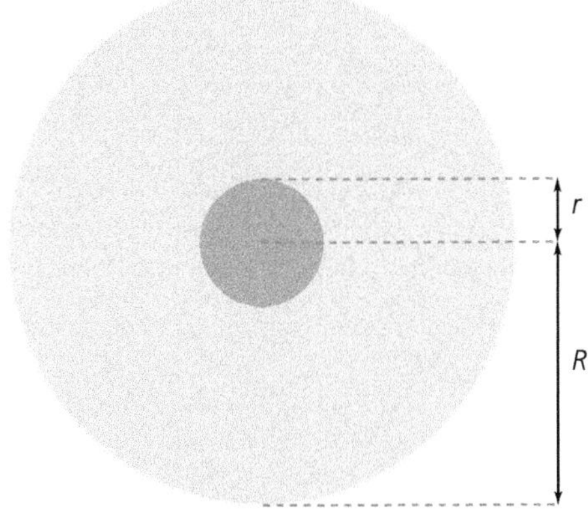

Mechanical advantage in pulleys

$$\text{Mechanical advantage (Pulley)} = \frac{\text{Radius of the wheel}}{\text{Radius of the axle}} = \frac{R}{r}$$

$$\text{Mechanical advantage (Levers)} = \frac{\text{Load}}{\text{Effort}}$$

The longer the lever arm, the easier it is to lift the object.

Levers:

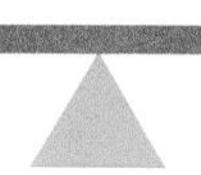

Hinge point

First class lever:

- Like a see-saw arrangement.
- Long arm for the force effort to push down on.
- Short arm for the load.
- Examples: float operated valve, claw hammer.

Balancing force: If the distance from the fulcrum or hinge point to the weight (10 kg) was 1 m and the distance from the fulcrum or hinge point to the force was 2 m you would only need 5 kg to balance the lever.

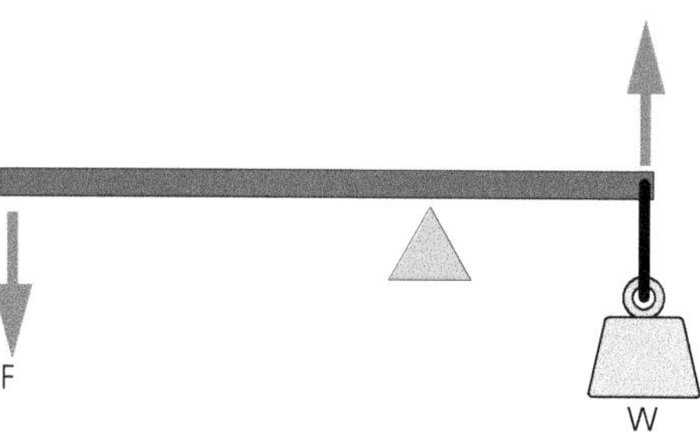

First class lever

Second class lever:

- Like a door, the fulcrum is at one end of the lever (upwards).
- The load is part way up the lifting arm.
- Examples: wheelbarrow, crowbar.

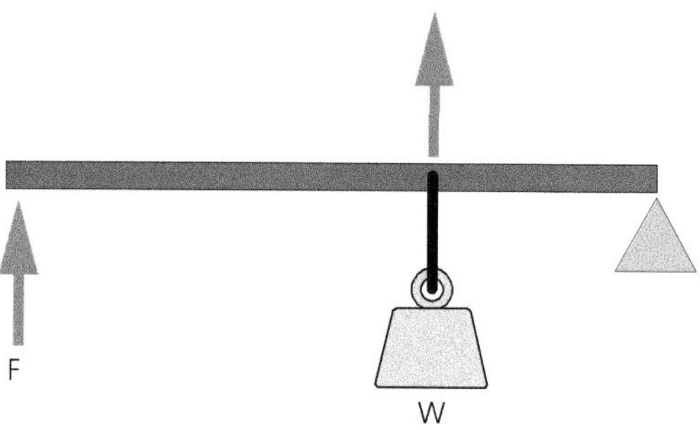

Second class lever

Third class lever:

- Like a door, the fulcrum is at one end of the lever (downwards).
- The lifting force is part way up the lifting arm, with the load at the opposite end to the fulcrum.
- Examples: human arm, spade.

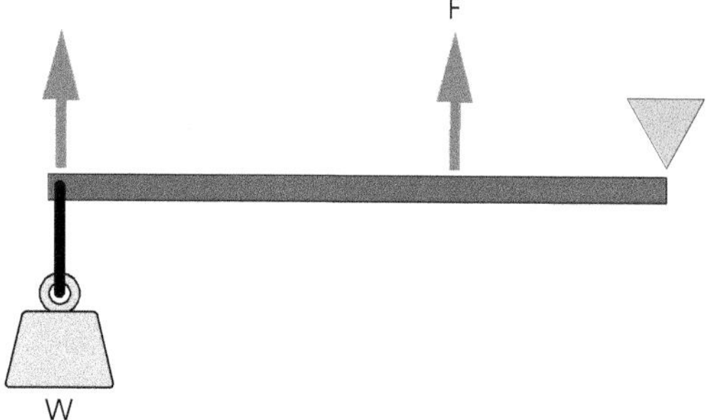

Third class lever

Exam tip

Take note of where the **fulcrum** is located, as this will help you identify the type of lever.

Fulcrum The hinge point for a lever.

Pulleys:

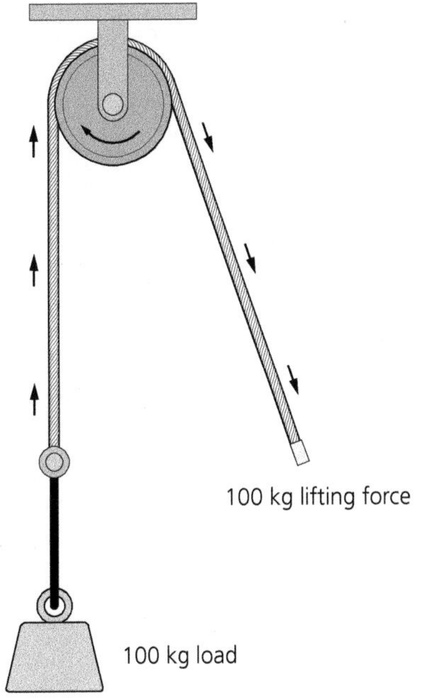

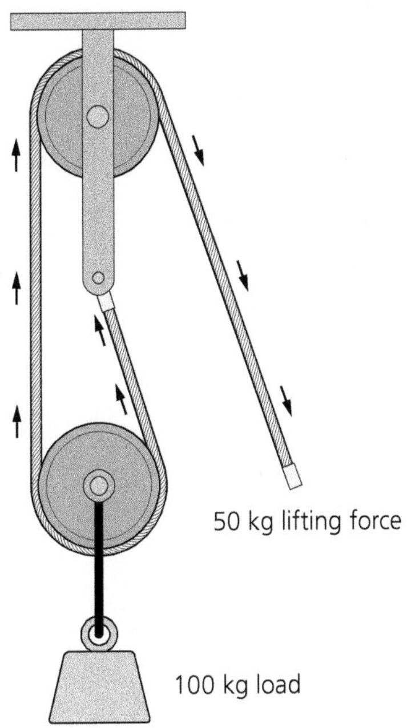

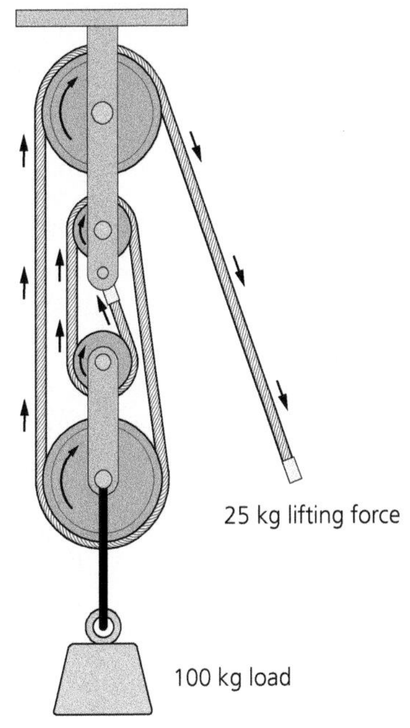

Pulleys

Looking at these different pulley systems, you divide the load to be lifted by the number of pulleys in the system to work out the advantage and therefore the lifting force required.

- The first system has one pulley, so 100 kg load divided by 1 = 100 kg lifting force.
- The second system has two pulleys, so 100 kg load divided by 2 = 50 kg lifting force.
- The third system has four pulleys, so 100 kg load divided by 4 = 25 kg lifting force.

Exam tip

Remember that the more pulleys, the easier the lift:

- one pulley = 100 kg lift
- two pulleys = 50 kg lift (but still raising the 100 kg); divide the load by the number of pulleys (mechanical advantage)
- four pulleys = 25 kg lift (but still raising the 100 kg).

Archimedes screw:

- This converts rotation into a straight line.
- A simple water pump.
- Similar to screws, threads and so on.
- Used to move volumes of water uphill or solid fuel to a burner.

LO4 Understand heat and power in the plumbing industry

Topic 4.1 Approaches to measuring temperature

REVISED

You can use the following units of measurement to measure temperature:

- Celsius (°C) – normal unit used by plumbers and customers.
- Kelvin (K) – scientific unit used to measure temperature.

The freezing point of water is 0°C = 273 K.

Absolute zero (where molecular motion stops) is –273°C = 0 K.

Devices can also be used to measure temperature, as outlined in Table 2.11.

Table 2.11 Devices used to measure temperature

Glass thermometer	+ Measures °C temperature of an object + Very common + Glass tube filled with mercury + As the temperature rises, so does the mercury level
Gas thermometer	+ Measures the variation in volume or gas pressure
Thermocouple	+ Two different metals are connected and when heated, they produce an electrical current + Senses temperature difference and therefore offers control + Older boilers used these to sense the pilot light being on
Thermistor	+ These are resistors where the resistance varies with temperature + Used on boiler high limit cut offs
Infrared thermometer	+ Detects temperatures using an infrared beam
Digital thermometer	+ Commonly used in plumbing industry + Often able to read two different temperatures, which is important when measuring flow and return

Check your understanding

10 Which type of thermometer would you use when reading the temperature of hot water coming out of a hot water tap?

Topic 4.2 Changes of state

REVISED

Materials will change state depending on the temperature in the environment. This not only happens in the environment (the rainwater cycle), but also in systems (condensing boiler).

Table 2.12 Changes of state

Melting	+ Ice to water (above 0°C) + Solid to liquid + Molecules begin to move apart
Freezing	+ Water to ice (0°C) + Liquid to solid + Molecules closely packed and stationary
Boiling	+ Water to steam (above 100°C) + Liquid to gas by increasing temperature + Molecules move far apart (expands 1600 times)
Evaporating	+ Liquid to gas by atmospheric conditions + Sun, light and wind as in the water cycle in Chapter 3 (Unit 214) + Molecules move far apart
Condensing	+ Steam to water + Gas to liquid (e.g. rain clouds forming or condensation on a cold mirror in a bathroom) + Molecules move closer together

Sensible heat:

- This means that as heat is applied, the temperature rises, so there is no change in state.
- For example, heating a hot water cylinder up means that the water stays as water.

Latent heat:

- Heat that causes any change in state is known as latent heat.
- For example, ice melting (solid to liquid); a kettle that keeps boiling (liquid to steam).

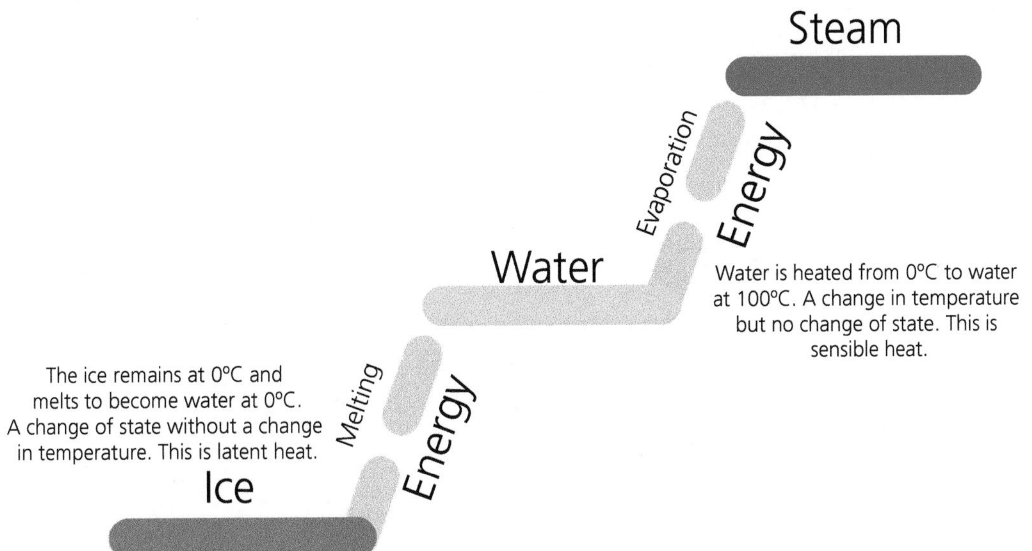

Sensible and latent heat

Check your understanding

11 The water in a hot water cylinder is heated from 15°C to 60°C. Is this sensible heat or latent heat?

Topic 4.3 Heat transfer

REVISED

Heat transfer is how quickly or slowly a material passes on heat to another material.

As plumbers, there are times we need heat to be transferred quickly, like a radiator heating a room or the coil heating the hot water in a cylinder. There are also times we do not want heat to be transferred – this is where we would use insulation.

Table 2.13 Methods of heat transfer

Conduction	• Heat travels through or along a substance • One molecule to the next molecule • The better the conductivity, the faster the heat will travel • Copper is a very good thermal conductor • If you hold a cold piece of copper pipe and put the other end of the pipe in a flame, the heat will travel up the pipe until it is too hot to hold

Convection	+ Only takes place in liquids and gases + Heated liquids or gases are less dense (so they will rise) + Cooler liquids or gases are denser (so they will fall, as in the old gravity boiler system below) 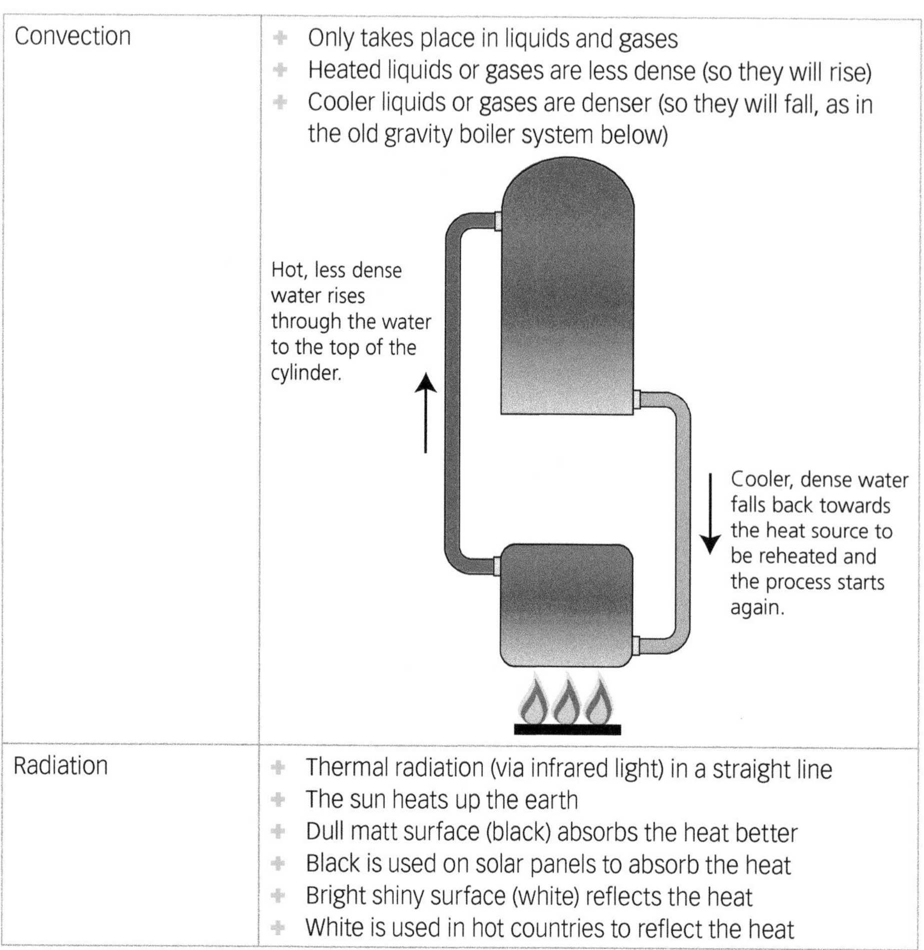
Radiation	+ Thermal radiation (via infrared light) in a straight line + The sun heats up the earth + Dull matt surface (black) absorbs the heat better + Black is used on solar panels to absorb the heat + Bright shiny surface (white) reflects the heat + White is used in hot countries to reflect the heat

Typical mistake

A question may ask about the main way a hot water cylinder is heated up. Think carefully: the coil heats the water in the cylinder, allowing convection to take place in the water.

Exam tip

In an exam, you may be asked about the main way that a room heats up from a radiator. A radiator only radiates about 15 percent of its heat. 85 percent of the heat is transferred by convection currents in the air.

Typical mistakes

Unable to recall the heat transfer methods and relating them to where they are found in systems. Try to remember the real-life situations where heat transfer happens in a system.

Topic 4.4 Units of energy and heat

REVISED

Energy and heat will use the following units of measurement:

Table 2.14 Units of energy and heat

Heat energy	Joule (J) or kilojoule (kJ)
Power	Watt (W) or kilowatt (kW)
Specific heat capacity	Kilojoules per kilogram per degree Celsius (kJ/kg/°C)
Coefficient of linear expansion (COLE)	Coefficient × material length × temperature rise = expansion (mm) Lead: 0.000029 Copper: 0.000016 Steel: 0.000011 Plastic: 0.00018

Worked example

You may get a question on temperature calculation, like the following:

How many kilojoules (heat energy) would it take to heat up 150 litres of water from 20°C to 60°C?

If you get a question like this, just multiply the three criteria you are given together! (You will need a calculator.)

Amount of water = 150 litres or 150 kg

Temperature rise = 60 – 20 = 40°C

Specific heat value of water = 4.186 kJ

The answer will be: 150 l × 40°C temperature rise × 4.186 kJ = 25,116 kJ

Check your understanding

12 How many kilojoules (heat energy) would it take to heat up 100 litres of water from 30°C to 80°C?

The specific heat capacity of water is 4.186 kJ. Calculate using: litre × temperature rise × specific heat capacity.

13 How many kilowatts (power) would it take to heat up 100 litres of water from 30°C to 80°C in one hour?

Calculate using: $\frac{\text{litre} \times \text{temperature rise} \times \text{specific heat capacity}}{\text{3600 (seconds in one hour)}}$

LO5 Understand the principles of electricity within the plumbing and heating industry

Topic 5.1 Principles of electricity; Topic 5.2 Units and formulae

REVISED 

- BS 7671 Design, installation and maintenance of domestic systems.
- Building Regulations part P – Domestic installations.
- Everything is made of molecules.
- Molecules are made up of atoms.
- Atoms have a nucleus.
- The nucleus is made up of protons (+ve), neutrons (no charge) and electrons (-ve).
- Electrons can pass from atom to atom.
- Materials that allow this free flow of electrons from atom to atom are **conductors**.
- Materials that do not allow this free flow of electrons from atom to atom are **insulators**.

Check your understanding

14 Name a good electrical conductor material.

15 Name a good electrical insulator material.

Table 2.15 Electrical measurements

Voltage	+ Also known as 'potential difference' + The greater the potential difference, the greater the pressure on the electron + Voltage (V) = Current (I) × resistance (R) + Domestic voltage 230 V + Safe site voltage 110 V
Resistance	+ The resistance to the movement of electrons through a conductor + If resistance is increased, current decreases + Resistance (R) = $\frac{\text{Voltage (V)}}{\text{Current (I)}}$ + Measured in Ohms + The symbol for ohms is Ω
Current/ Amperage	+ The rate at which electricity flows to an appliance + Measured in 'Amps' (I) + Fuses and MCBs are rated in amps + Current (I) = $\frac{\text{Voltage (V)}}{\text{Resistance (R)}}$
Power	+ Rate at which energy is converted into another form of energy, like heat, light and movement + For example, the power of an electric shower describes the rate at which electricity is converted into heat. So, a 10 kW electric shower would heat up water more quickly than an 8 kW electric shower.

MCB Micro Circuit Breaker or Mini Circuit Breaker.

Types of electrical current

Direct current (DC):

- The electrons flow from negative to positive.
- Produced by a battery/cells, electrochemically.
- Flows in a simple direct current circuit, as in a torch.
- Symbol:

Alternating current (AC):

- The electrons reverse direction constantly (50 times per second in the UK, a frequency of 50 Hz).
- AC voltages can be changed using a transformer (step-up or step-down).
- Symbol:

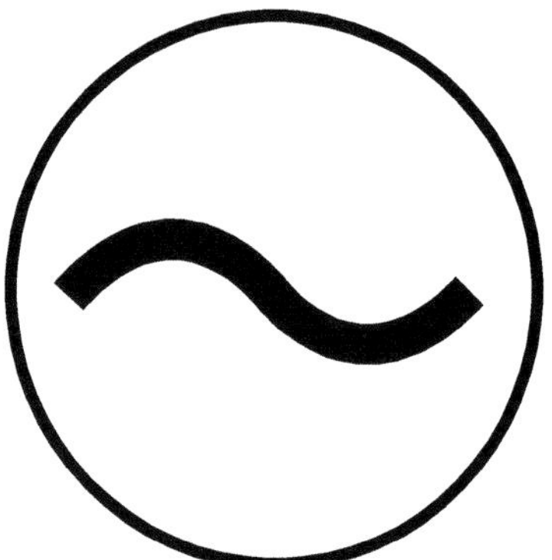

- Produced electromagnetically at a power station.
- Magnet + coil + movement = electricity

- Produces a sinusoidal wave, as shown in the diagram:

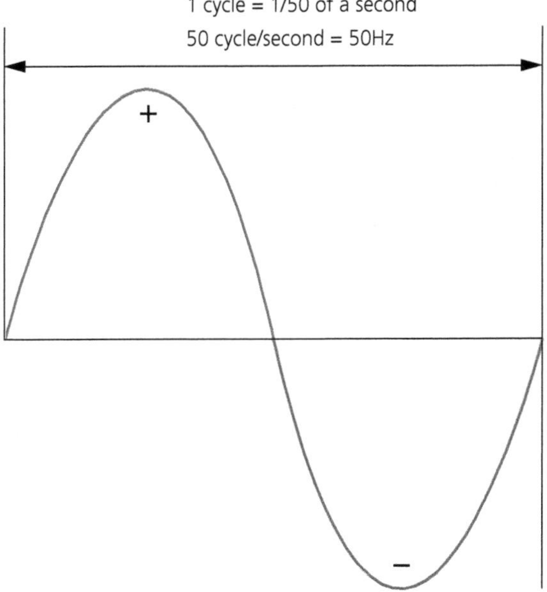

Power consumption of electrical units

The power triangle shows the relationship between power (P), current (I) and voltage (V).

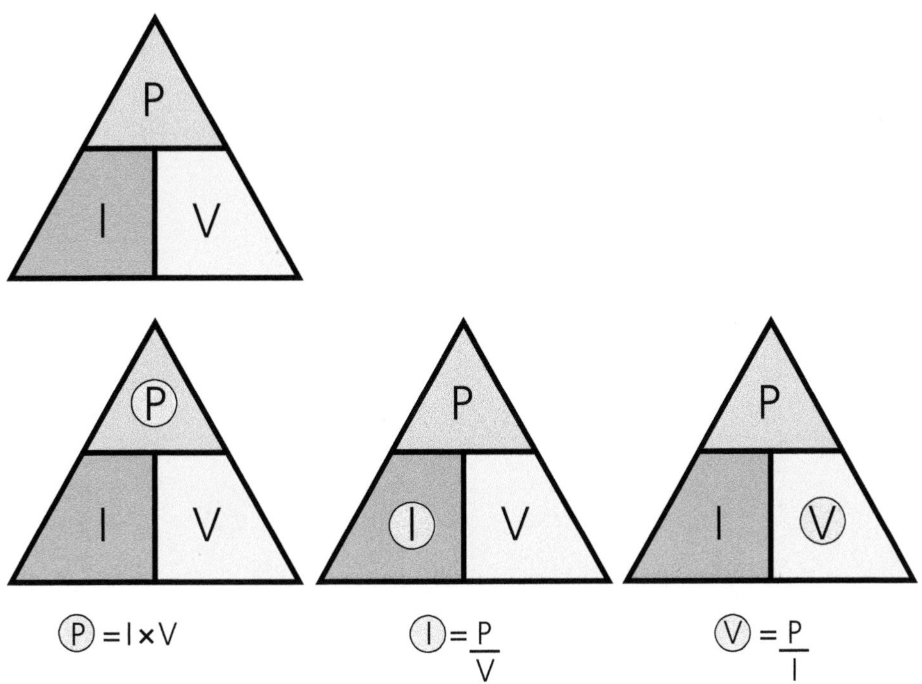

The triangle is used like this:

Whichever value you are trying to find, you cover with your finger. That leaves two criteria. If these two criteria are above each other you divide, if they are beside each other you multiply.

1. You want to find Power (watts). Cover P with your finger. That leaves current (I) and voltage (V) next to each other, so:
   Power (watts) = Current (I) × Voltage (V)
2. You want to find Current (I). Cover (I) with your finger. That leaves Power (P) above Voltage (V), so:

   $$\text{Current (I)} = \frac{\text{Power (P)}}{\text{Volts (V)}}$$

Exam tip

Remember that the size of a fuse or MCB is measured in amps (I), so if a question asks you to work out a fuse size, this is the equation you use!

3 You want to find Volts (V). Cover (V) with your finger. That leaves Power (P) above Current (I), so:

$$\text{Volts (V)} = \frac{\text{Power (P)}}{\text{Current (I)}}$$

Ohm's law

'The current through a conductor between two points is equal to the voltage across the two points and inversely proportional to the resistance between them.'

This triangle works in the same way as the power triangle did:

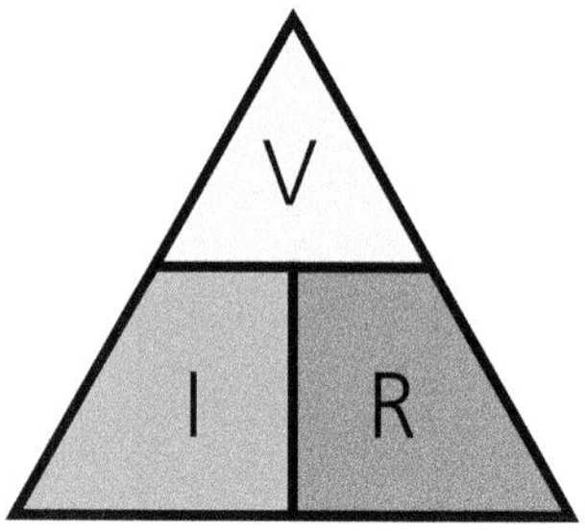

V
I R

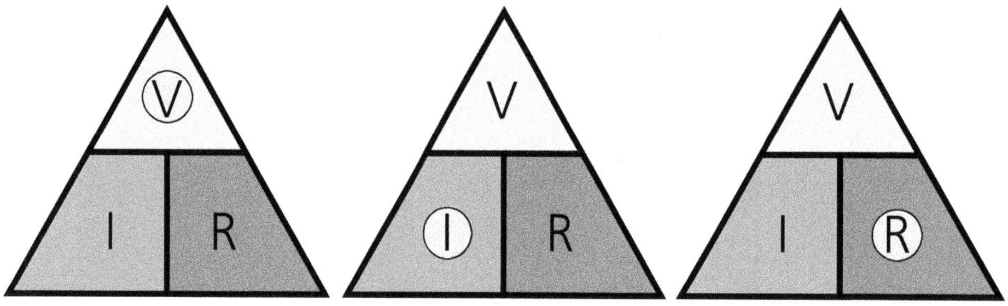

V I R
V I R
V I R

Ⓥ = I × R

$Ⓘ = \frac{V}{R}$

$Ⓡ = \frac{V}{I}$

1 Voltage (V) = Current (I) × Resistance (R)

2 $\text{Current (I)} = \frac{\text{Voltage (V)}}{\text{Resistance (R)}}$

3 $\text{Resistance (R)} = \frac{\text{Voltage (V)}}{\text{Current (I)}}$

Check your understanding

16 What size over-current protection would be needed to protect a circuit with a 3 kW immersion heater installed in a domestic property? Use the power triangle to work the size out.

17 Calculate the voltage in a circuit that has a resistance of 115 Ω, with a current of 2 A.

Typical mistakes

Students unable to recall the power triangle and Ohm's law triangle correctly. Make sure you can remember both the power triangle and Ohm's law triangle.

Exam tip

If the phrase 'overcurrent protection' is used in a question, it is referring to the fuse or MCB size.

Exam tip

When you get into the exam, write these two triangles down straight away so you don't forget them!

Exam tip

Question 16 has several parts.

1 You are asked to work out the over-current device size.
2 You have to remember that the over-current device is the fuse/MCB size (I).
3 You have to recall that domestic voltage is 230 V.
4 You have to recall the equation.

Table 2.16 Type of circuits

Series circuit	There is only one path from the source through all the loads and back to the source If a second load is added to the circuit, the resistance doubles (40 Ω to 80 Ω) and the current flow is halved, which means the bulbs are not as bright	40 Ω Source 230 V Switch Earth wire omitted for clarity R1 40 Ω R2 40 Ω Source 230 volt Switch Earth wire omitted for clarity
Parallel circuit	There are independent (parallel) circuits from and back to the source The bulbs are unaffected by each other. However, the amperage increases as more bulbs are added. Conversely, the resistance decreases	Source 230 V Earth wire omitted for clarity R1 40 Ω R2 60 Ω R3 100 Ω Switch Switch Switch

Topic 5.3 Circuit protection and earthing

Electrical protection components break circuits that are at risk of, or could cause, danger.

Table 2.17 Protection components

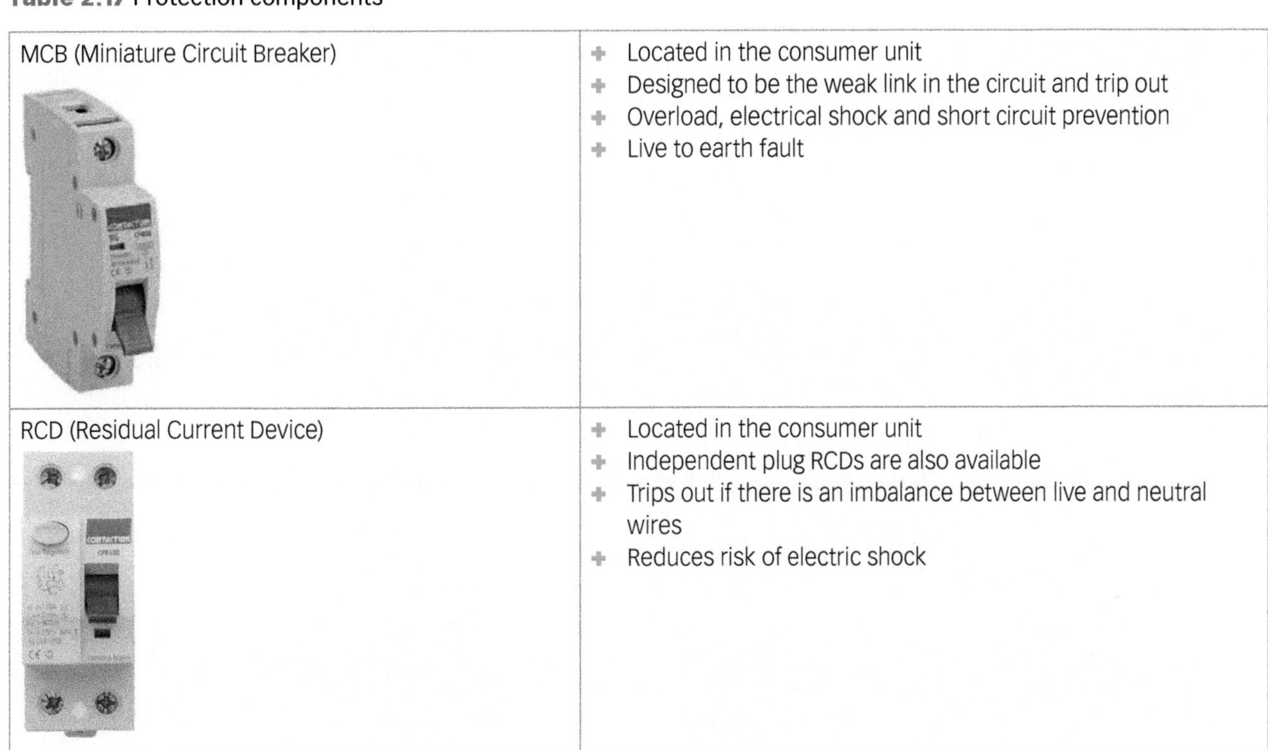

MCB (Miniature Circuit Breaker)	+ Located in the consumer unit + Designed to be the weak link in the circuit and trip out + Overload, electrical shock and short circuit prevention + Live to earth fault
RCD (Residual Current Device)	+ Located in the consumer unit + Independent plug RCDs are also available + Trips out if there is an imbalance between live and neutral wires + Reduces risk of electric shock

Cartridge fuse	+ Located in plug tops + The fuse in the plug top will protect the cable and appliance from being overloaded or burnt
RCBO (Residual Current Breaker with Over-Current)	+ Located in the consumer unit + Combines both MCB and RCD in one unit + Offers protection against overload and electric shock
Rewireable fuse 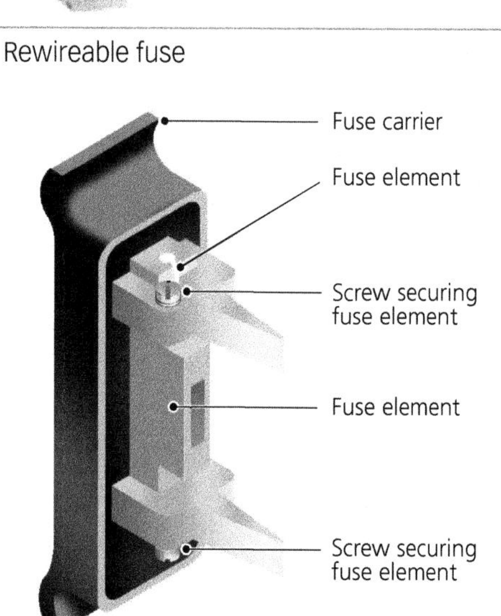 	+ Located in old consumer units (replaced by MCB) + The diameter of the wire relates to the size of protection + The wire will melt if overloaded, therefore breaking the circuit

Table 2.18 Protective equipotential bonding

<table>
<tr><td>Equipotential bonding

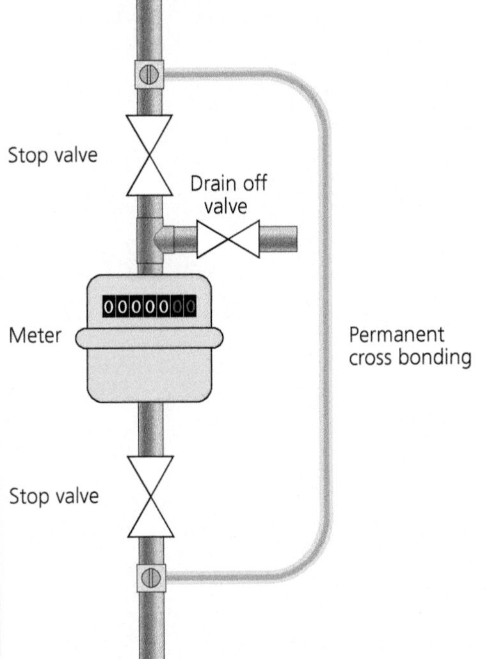
</td><td>
<ul>
<li>This connects (earths) all metal fixtures back to the main earth bar in the consumer unit</li>
<li>Pipework (hot, cold, gas, central heating), radiators, sinks etc.</li>
<li>10 mm³ earth cable (green and yellow) is clamped to each fixture</li>
<li>A second supplementary bond (6 mm³) can be made locally, linking all pipework together under an appliance, bath, sink or combi boiler</li>
<li>A plastic component, such as a plastic water meter, must be cross-bonded to maintain the circuit</li>
</ul></td></tr>
<tr><td>Temporary continuity bonding
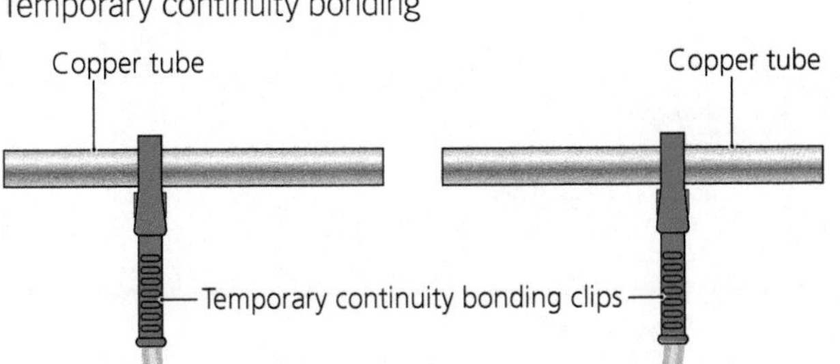
</td><td>
<ul>
<li>Used when cutting into a piece of pipework to maintain the circuit</li>
<li>10 mm³ earth cable with two crocodile clips</li>
<li>Installed before the pipework is cut to prevent electric shock</li>
</ul></td></tr>
</table>

Exam-style questions

1 Which way is the force of a liquid exerted?
- a Downwards only
- b Sideways only
- c Downwards and sideways
- d Upwards and downwards

2 What is potential difference measured in?
- a Volts
- c Ohms
- b Amps
- d Watts

3 Which of the following installations offers the least resistance to the flow of water?
- a Four elbows and two machine bends
- b Two elbows and four machine bends
- c Five elbows and one machine bend
- d Six elbows

4 Which voltage is domestic voltage?
- a 110 V
- c 400 V
- b 230 V
- d 18 V

5 When cutting into pipework to add a connection, what type of earth bonding should be put in place to keep the operative safe?
- a Supplementary bonding
- b Equipotential bonding
- c Cross bonding
- d Temporary continuity bonding

6 What is the SI unit for temperature measurement?
- a Kelvin
- c Joules
- b Fahrenheit
- d Pascal

7 An old direct hot water cylinder is attached to a boiler with no pump. How is the heat transferred to the hot water cylinder?
- a Conduction
- c Impulse
- b Convection
- d Radiation

8 What protective device is positioned in the consumer unit and protects a whole circuit?
- a Cartridge fuse
- c MCB
- b RCD
- d Resetting plug top

9 In a hot water system, there are components made of dissimilar metals that are connected together. Which one of the components destroys the other due to electrolytic corrosion?
- a Inhibitor destroys the cathode
- b Anode destroys the cathode
- c Cathode destroys the anode
- d Rust destroys the anode

10 Using the power triangle, what is the fuse rating for a circuit with 230 V and a power rating of 690 W?
- a 13 A
- c 5 A
- b 10 A
- d 3 A

11 What happens to a gas, which has a relative density of greater than 1, when left in air?
- a It rises
- c It burns better
- b It falls
- d It burns worse

12 At what temperature is water most dense?
- a 0°C
- c 10°C
- b 4°C
- d 100°C

13 By approximately what percentage does water expand by when heated between 10°C and 95°C?
- a 4%
- c 25%
- b 10%
- d 1600 times

14 What element must be present to form a ferrous metal?
- a Copper
- c Iron
- b Brass
- d Aluminium

15 Which of the following metals is not an alloy?
- a Brass
- c Low carbon steel
- b Bronze
- d Copper

16 Which phrase best describes a parallel electrical circuit?
- a The more bulbs added to the circuit the dimmer they become
- b Each bulb is unaffected by the addition of more
- c It is one big continuous circuit
- d The more bulbs added to the circuit the brighter they become

17 Which of the following outlets has the greatest pressure?

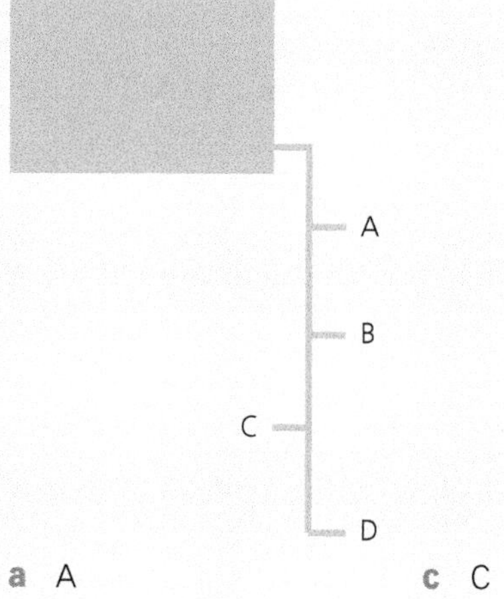

- a A
- c C
- b B
- d D

18 What does this symbol mean?

- a AC electricity
- c Water meter
- b DC electricity
- d Flow of water

19 Which of the following materials has the best ductility?
- a Brass
- c Low carbon steel
- b Bronze
- d Copper

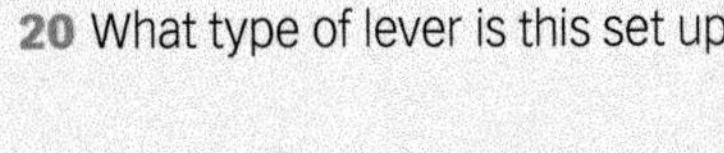

20 What type of lever is this set up?

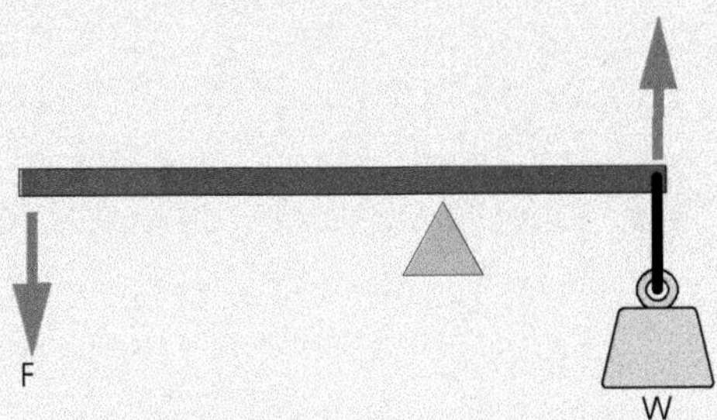

- **a** First class lever
- **b** Second class lever
- **c** Third class lever
- **d** Fourth class lever

21 Which is the correct equation to find resistance?

- **a** Voltage / Current
- **b** Current / Voltage
- **c** Power / Current
- **d** Voltage / Power

22 How many kilojoules of heat energy would it take to heat up 100 litres of water from 20°C to 70°C? (Specific heat capacity of water is 4.186 kJ.)

- **a** 20,930 kJ
- **b** 29,302 kJ
- **c** 8372 kJ
- **d** 1194 kJ

23 In a pulley system, how do you work out the required lifting force?

- **a** Multiply the weight to be lifted by the number of ropes
- **b** Multiply the weight to be lifted by the number of pulleys
- **c** Divide the weight to be lifted by the number of ropes
- **d** Divide the weight to the lifted by the number of pulleys

24 What makes a good insulator?

- **a** The electrons can flow
- **b** The electrons cannot flow
- **c** The neutrons can flow
- **d** The neutrons cannot flow

25 Which of the following processes shows sensible heat?

- **a** Steam from a kettle when boiled
- **b** Water dripping off an ice cube as it thaws
- **c** Hot water heating up in a hot water cylinder
- **d** Condensation forming on a cold surface

3 Cold water (Unit 214)

You need to familiarise yourself with the pipework, systems, appliances and components in cold water systems and know how to preserve the quality of water within the systems, along with the maintenance requirements. Ask yourself:

- Where does water come from?
- How is water quality maintained?
- How do systems work and how are they maintained?
- What documentation is there for cold water systems?

There are some practical workshop activities which will need to be completed as part of this unit, mainly in Learning Outcome 3. The concepts explained and undertaken in these activities will also help you in the exam.

LO1 Understand cold water supply to dwellings

Topic 1.1 Sources and properties of water

REVISED

The rainwater cycle is a natural process where water is continually exchanged between the atmosphere, surface/ground water and land.

- The sun warms the earth and some of the surface water evaporates.
- The vapour rises and condenses forming clouds.
- Clouds become saturated and it rains (precipitates).
- Water reaches the ground and goes into rivers, lakes and seas.

Evaporate When water molecules move apart and turn to gas.

Condense When molecules move together and form water droplets.

Precipitation Water that falls from clouds to the ground, such as rain, snow, sleet.

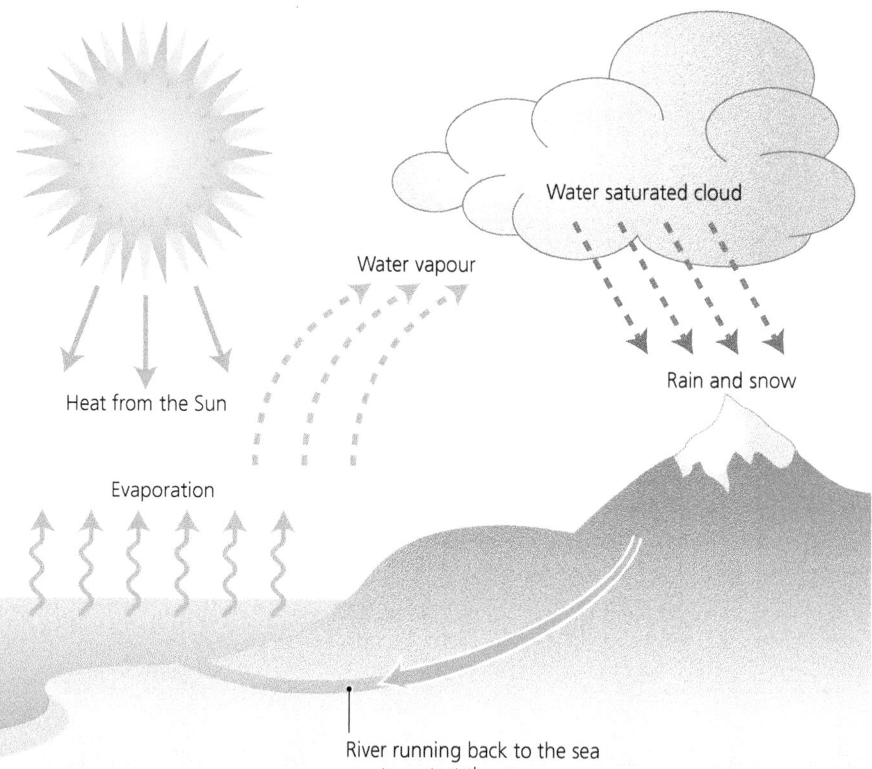

The rainwater cycle

Water sources

There are **surface sources** (for example, lakes, rivers, canals, reservoirs and streams) and **underground sources** (such as wells, boreholes and springs). The quality of water direct from these sources tends to be poor due to contamination, so water treatment is necessary.

There are different categories of water to inform us of the various contamination levels there are within properties. Always prevent the contamination of Category 1 water.

Fluid categories of water

Table 3.1 Fluid categories of water

Category	Description
Category 1	+ Wholesome, potable drinking water + Supplied by the water undertaker direct to domestic properties + Complies with water quality regulations
Category 2	Category 1 water BUT aesthetic quality has been altered due to: + temperature + taste + colour + smell Similar to making a cup of tea!
Category 3	**Slight** health risk Not suitable for drinking Examples: + primary circuit in a heating system + soapy water from a bath or basin + washing machine water
Category 4	**Significant** health risk Not suitable for drinking Strong chemicals could be present Examples: + swimming pools + garden hoses + commercial primary circuits + circuits with anti-freeze (solar)
Category 5	**Severe** health risk Not suitable for drinking Examples: + foul water (e.g. WC, urinals) + grey water + laboratories, hospitals + slaughterhouses, agricultural systems

Exam tip

Remember that wholesome water, potable water and drinking water are the same thing.

Exam tip

Remember the differences between the fluid categories for the exam: 'slight' is 3, 'significant' is 4 and 'severe' is 5.

These categories of water are listed in the Water Supply (Water Fittings) Regulations as well as BS EN 806.

It is imperative when installing systems that cross connections between water categories are not made, as this would be a form of contamination

Typical mistakes

Not being able to relate the categories of water to situations within a domestic property. It is important to know this so you can identify the correct back flow protection.

Check your understanding

1 Which fluid category would water that has got hand soap in it be classified as?

- **Category 1 water**: wholesome water supplied by the water undertaker. Used for drinking, food preparation and washing.
- **Recycled greywater**: collected from bath, basin and shower waste water. Used to flush toilets and washing machines.
- **Rainwater harvesting**: collected from the roof line and used for flushing toilets, supplying outside taps and washing machines.

Topic 1.2 Types of supply into a property

REVISED

The types of supply into a property include:

- Undertaker's main – wholesome, potable drinking water, Category 1, supplied by the water undertaker. Covered by the Water Supply (Water Fitting) Regulations 1999.
- Private source – water from a well or borehole so not connected to the water mains. Covered by the Private Water Supply Regulations 2016.
- Rainwater harvesting – collected from roof lines, filtered and stored. Category 5 water, then used to flush toilets and supply outside taps.
- Recycled water – collected from bath, basin and shower wastes, referred to as greywater. Category 5 water, then used to flush toilets and supply washing machines.

Check your understanding

2 Along with the Water Regulations, what British Standard lists the categories of water?

The Water Supply (Water Fittings) Regulations 1999 require wholesome water to be provided with systems to prevent the following problems:

Table 3.2 Problems with the supply of wholesome water

Waste of water	+ Caused by leaks and drips, faulty installations, poor maintenance
Undue consumption	+ Make sure appliances do not use more water than necessary + Install showers, rather than baths + Install dual flush WCs
Misuse	+ Domestic and commercial water rates must be correctly imposed
Contamination	+ Category 1 water must ALWAYS be protected + Install systems to prevent backflow/back siphonage and cross-flow
Erroneous measurement	+ Water entering a house should go through a meter + Pay for what you use + Don't bypass the meter

Exam tip

Contamination is the big one to avoid in systems. Protecting wholesome, potable water is important, so there will be questions on how to protect Class 1 water from the other classes, by both mechanical and non-mechanical means.

Topic 1.3 Treatment and distribution of cold water

REVISED

Treatment methods of mains cold water include:

Table 3.3 Treatment methods of mains cold water

Sedimentation	+ These tanks slow down the velocity of water and allow the suspended large solids to sink under gravity + The inlet and outlets are on opposite ends + The tanks are covered to avoid additional contamination + The tanks must be regularly cleaned

Filtration	+ This process removes smaller particles and algae
	+ Slow sand filters:
	  + Water slowly soaks through a layer of biological sludge or 'schmutzdecke', along with fine sand
	  + The schmutzdecke is removed when the flow of water is low
	+ Rapid filters:
	  + Water is slowly forced through fine sand and gravel which filters particles out
	  + The sand is back washed when the flow of water is low
	+ Pressure filters:
	  + Water is forced under pressure through sand and gravel filters
	  + The sand is back washed when required

Sterilisation	+ After filtration, the water is sterilised before entering the mains system + Chlorine and ammonia are added to kill off any remaining bacteria + Ultraviolet (UV) treatment can also be used, which kills off remaining bacteria and viruses

Check your understanding

3 The Water Regulations list certain problems that must be prevented when installing a cold water system in a property. Contamination is the major area.
The regulations also list undue consumption and misuse. What is another problem area?

There are also distribution treatment methods of mains cold water, which include:

Gravity:

+ Works on head height.
+ Reservoirs are at high level.
+ Fed into slow sand filters.
+ Chlorinated before distribution.
+ Homes are below reservoir height.

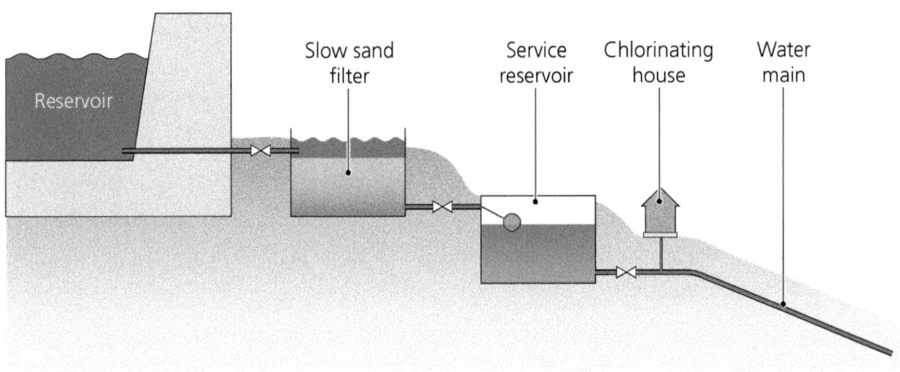

Gravity distribution

Pumped:

+ Water is pumped through the system.
+ Slow sand filters are used.
+ Water is pumped up to head height (water tower).
+ Water towers have now been replaced with large pumps that distribute the water around the mains.

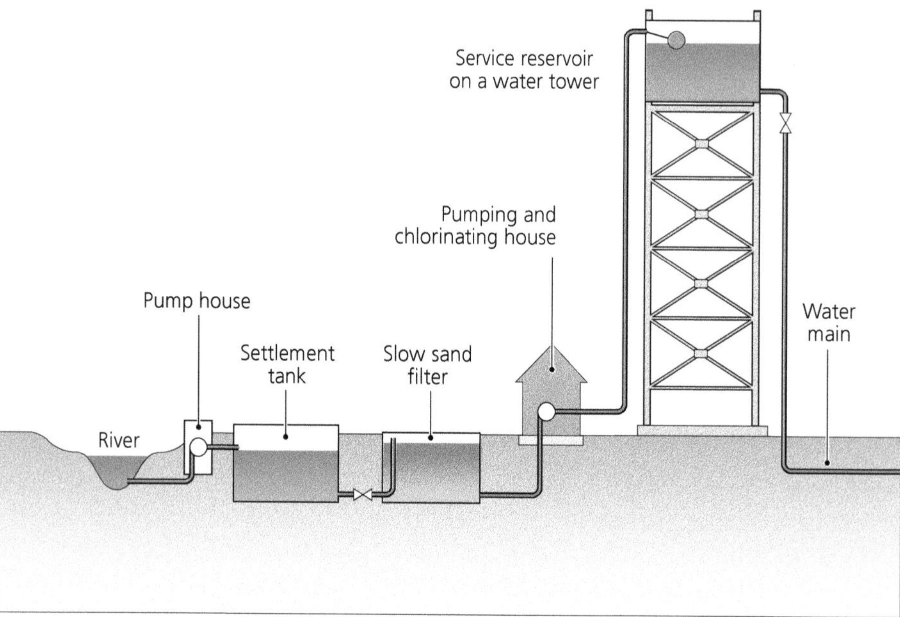

Pumped distribution

Check your understanding

4 What does adding a small amount of chlorine to the treated water at the final stage do?

5 Which part of the water treatment process removes any remaining bacteria from the water?

Now test yourself TESTED

1 Explain how cross-flow contamination could occur in a mixer valve in a shower.

LO2 Understand domestic cold water systems

Topic 2.1 Sources of information relating to systems

REVISED 

You must be aware of the sources of information that must be used when undertaking work on cold water systems. These include:

Statutory regulations:

- The Water Supply (Water Fittings) Regulations 1999.
- Replaced all the by-laws.
- Waste of water / undue consumption / misuse / contamination / erroneous measurement.
- Cover water supplied by the water undertaker.
- Only approved materials can be used – lead must not be used.

The Private Water Supplies Regulations 2016:

- Cover water not supplied by the water undertaker.
- Supplied from a borehole, spring, well, river, stream or pond.

Industrial standards:

- BS EN 806 – design, installation, commissioning, testing, flushing and disinfecting of domestic systems. Used with BS 8558.
- BS 8558 – guide to design, installation, testing and maintenance. Used in conjunction with BS EN 806.
- (BS 6700 – although this has been withdrawn, it is still referenced where information is not available.)

Manufacturers' technical instructions:

- These are key when installing appliances and equipment and include:
  - installation instructions
  - testing and commissioning procedures
  - information on spare parts
  - fault finding.

Topic 2.2 Service pipework layout

REVISED

This outlines key information regarding the incoming cold water main, from the water main in the road up to the rising main in the property.

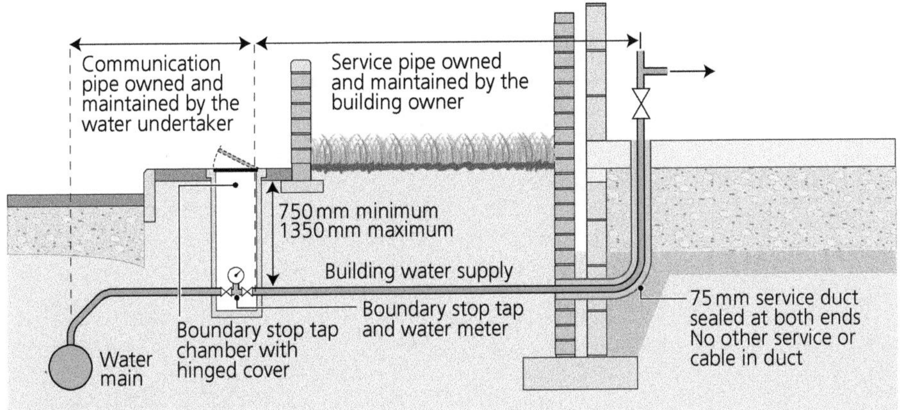

Domestic water supply

> **Exam tip**
>
> This is an important diagram to understand as you might be asked many different questions on the detail that it shows, such as names of pipes, depths, insulation, the isolation point and where the meter is located.

- Connected to the mains via a **ferrule**.
- **Communication pipe** runs from the mains to the meter.
- **Service pipe** runs from the meter to the house.
- **MDPE** (medium density polyethylene) **BLUE** pipe underground.
- Underground between a depth of **750 mm to 1350 mm** – to protect from **frost**.
- **External stop valve** in meter box.
- **Internal stop valve** – as near as possible to the entry to the building.
- MDPE pipe is **insulated in a duct** as it enters the property.
- Duct is sealed at both ends to prevent rodent entry.
- The MDPE pipe should change to copper (about 150 mm above ground level).
- Where the copper pipe starts, there should be the main internal isolation point (stop valve).
- There should be a drain off valve immediately above the stop valve.
- Above the drain off, the pipework becomes the rising main.

> **Typical mistakes**
>
> Not knowing the details from the drawing of the incoming mains well enough. You must remember the dimensions, names and design.

> **Check your understanding**
>
> 6 What is the primary reason the service pipe has to be laid between 750–1350 mm deep?

Table 3.4 Supply pipe materials

Medium density polyethylene (MDPE)	+ Coloured blue to identify wholesome water + Must be underground or pipe will corrode + Most common modern material for supply pipework
Copper	+ R220 soft copper + This allows for ground movement
Lead	+ Older houses may still have a lead supply + Propriety fittings used only + Must not install new lead pipework – The Water Supply (Water Fittings) Regulations and BS EN 806

Check your understanding

7 What is the name of the fitting that makes the connection onto the water main in the road and allows a feed to be made to a property?

Now test yourself

2 At a customer's property, you need to isolate the incoming mains but the isolation stop valve under the kitchen sink is seized. Where else can you isolate the incoming mains while you change the stop valve?

Topic 2.3 Types of cold water systems

REVISED

Direct system:

+ outlets direct from the mains
+ wholesome water to all outlets
+ no stored water
+ high pressure to all outlets
+ 15 mm pipework.

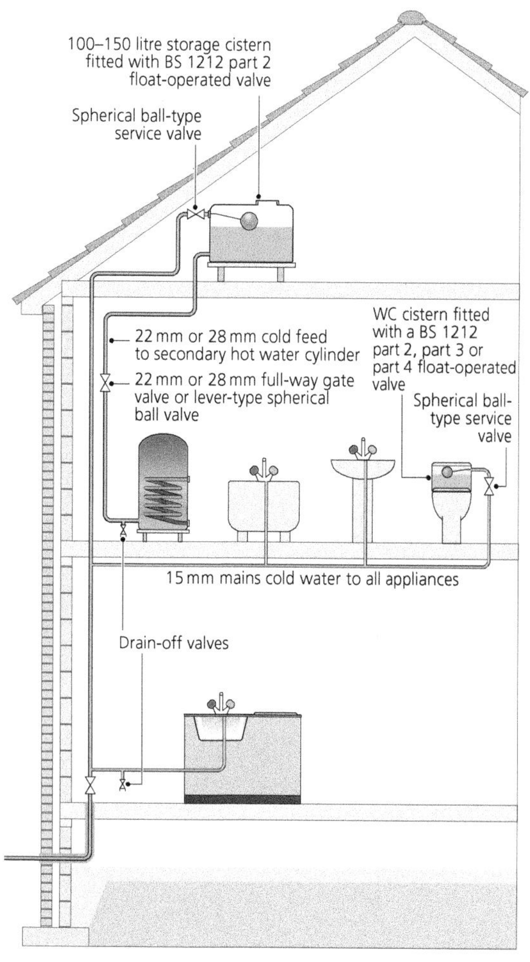

Direct cold water, indirect hot water

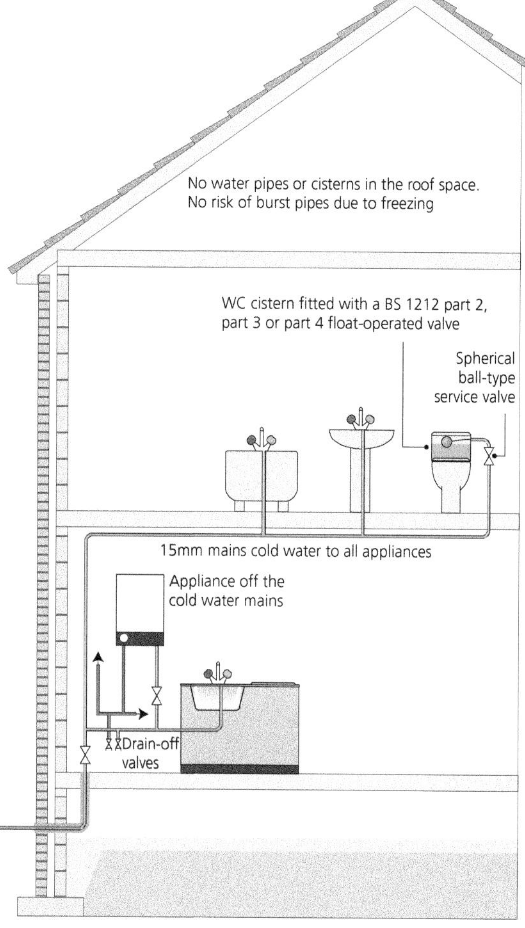

Direct cold and hot water

Indirect system:

- only kitchen sink fed from mains
- other outlets fed from cistern
- stored cold water
- larger pipework
- low pressure.

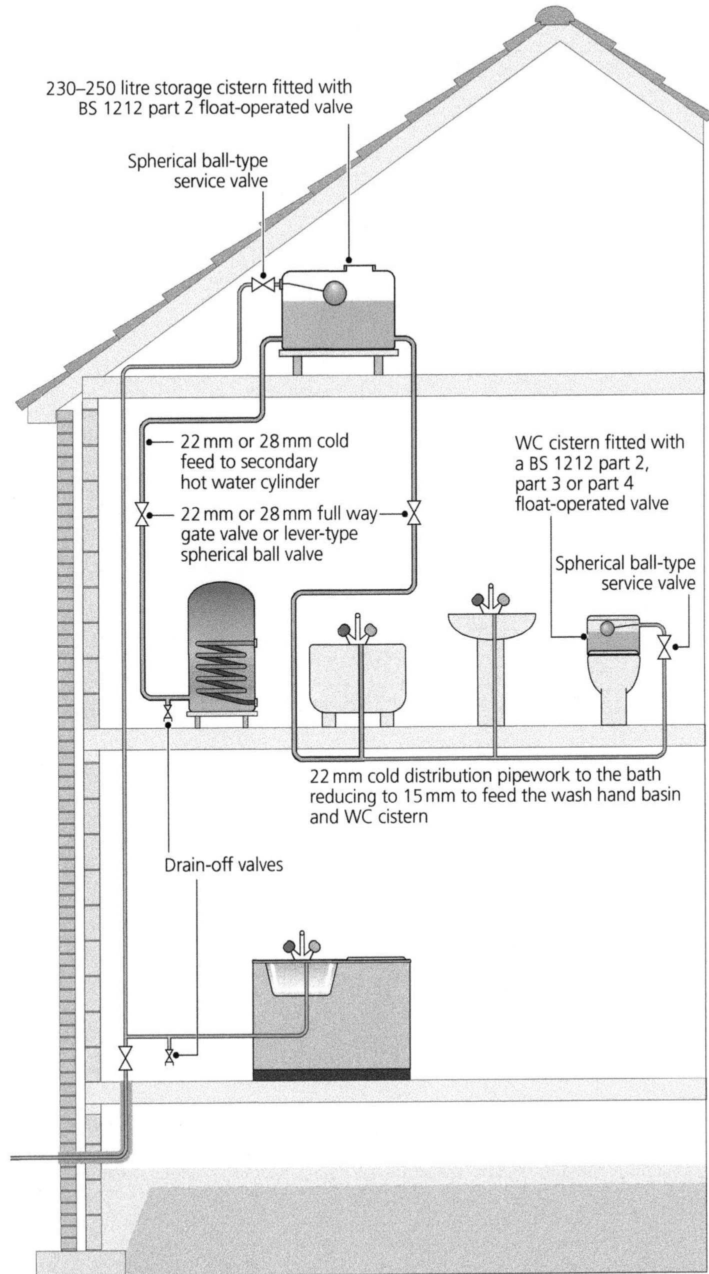

Indirect system

Table 3.5 Advantages and disadvantages of direct and indirect systems

Direct system	**Advantages** + Cheap to install + Wholesome water at outlets + Less pipework + Suitable for showers, taps and mixers + Smaller pipework diameter + Higher pressure at outlets	**Disadvantages** + Pressure could vary during the day + Mains under repair – no water + Noise + Risk of contamination + Water hammer + Condensation on pipework
Indirect system	**Advantages** + Reduced risk of water hammer + Constant low pressure + Suitable for vented system + Reserve if mains supply is cut off + Pipework sized for greater flow rates	**Disadvantages** + Frost protection required + Back flow protection required + Space taken up by cistern + Increased cost of installation + Low pressure

Installation requirements

The Water Supply (Water Fittings) Regulations and BS EN 806 require the following:

Table 3.6 Installation requirements

Cistern size	+ Should be capable of supplying wholesome water (24 hrs)
	+ Sized correctly
	+ BS EN 806:
	  + Small house
	    – Cold water only 100–150 litres
	    – Hot water only at least the capacity of the hot water cylinder
	    – Hot and cold water 230 litres
	  + Large house
	    – 100 litres per bedroom

Warning pipe	+ One size bigger than the inlet + Located to indicate problem + Must have screen to prevent insects being able to get in
Inlet and outlet position	+ Float operated valve (inlet) should be located opposite the outlet to avoid stagnation
Position of float operated valve	+ BS 1212 Part 2 (brass) or Part 3 valve (plastic) + Outlet above the body to allow for air gap
Position of cistern vent	+ Located in the lid + Must have a screened vent to avoid insects being able to get in
Position of open vent pipe connection	+ Open vent from the hot water cylinder must be located over the cold water cistern + It must have a grommet between the pipe and rigid close fitting cistern lid to avoid insects being able to get in
Cistern lid	+ Rigid close-fitting lid with a snap + Prevent insects, debris and light getting to the water + There must be 350 mm minimum above the lid to allow for maintenance
Service valve	+ Prior to every float operated valve there must be a service valve (high and low pressure service valves have different bores) + Allows maintenance and replacement
Insect screen	+ Mesh which prevents insect access (contamination) + Located on warning pipe and vent
Insulation	+ Prevents frost damage in the winter and heat damage in the summer + Located around cistern and on top of cistern + The ceiling insulation should be removed from under the cistern to allow a small amount of heat to the cistern from the room below
Support	+ Should be located over a load bearing wall or a minimum of three joists + The cistern must be supported over its whole base
Drilling	+ When positioning a tank connector or drilling a hole for the warning pipe, float operated valve or vent, a hole saw MUST be used
Linking cisterns	+ Larger properties with greater demand may require cisterns to be connected/linked giving greater capacity of stored water + Stagnation MUST be prevented by having the inlet in one cistern and the outlet in the second cistern. This allows water to flow through
Legionella control	+ Legionella is in all water. We must not allow it to grow, as legionella can cause major health issues in humans + Regulations state we must not allow the cold water cistern to get to 20°C, so therefore the insulation must prevent heating of the water in summer + Cisterns must be sized correctly so the capacity of the cistern is used within 24 hrs

Grommet A rubber seal used on the cold water storage cistern.

Stagnation When water is allowed to stand still and becomes stale and foul.

Check your understanding

8 What is likely to cause stagnation in a cold water storage cistern?

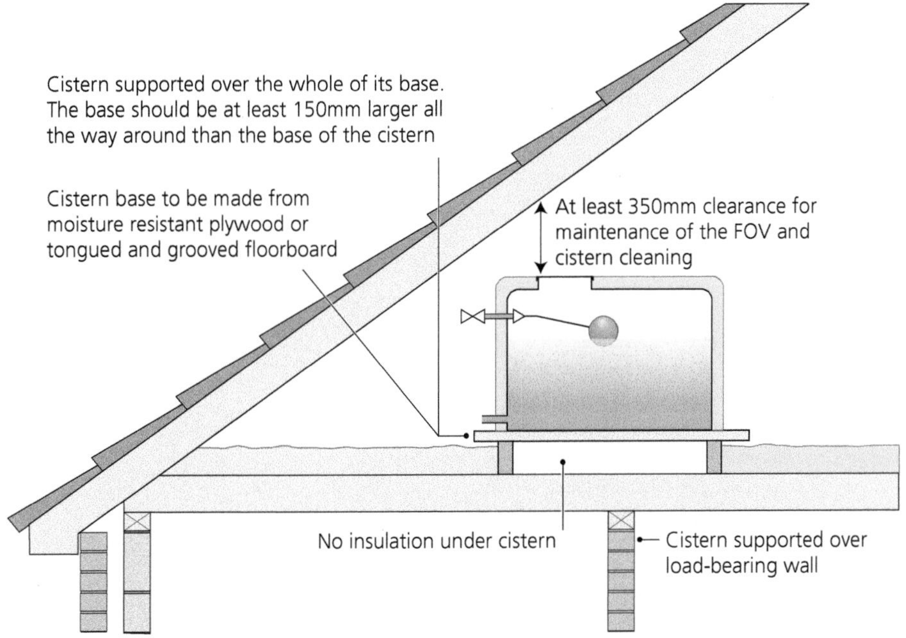

Cold water storage cistern

Typical mistakes

Unable to recall installation details for the cold water storage cistern. Try and relate all the detail in the table above to this diagram as well as the components outlined later in this chapter.

Topic 2.4 Operating principles of backflow prevention devices

REVISED 

- Backflow is water flowing in the wrong direction due to loss of pressure.
- Back siphonage is a vacuum which sucks water backwards due to a change in pressure.

You need to know the operating principles and uses for basic types of backflow prevention devices, including mechanical and non-mechanical devices.

Mechanical devices

Single check valve:

- Use to protect against hot water, like shower valves or mixer taps.
- Non return valve.
- Protects Category 1 against Category 2.

Double check valve:

- Use to protect against mild chemicals, like filling loop for a sealed central heating system, an outside tap or a basin with no air gap.
- Protects Category 1 against Category 3.

Non-mechanical devices

Air gaps:

- AG
  - The air gap between the outlet of the float operated valve and the water level in the cistern.
  - Protects Category 1 against Category 3.
- AUK1
  - The air gap between the over-flow of a WC cistern and the spill over level of the WC pan.
  - Protects Category 1 against Category 5.

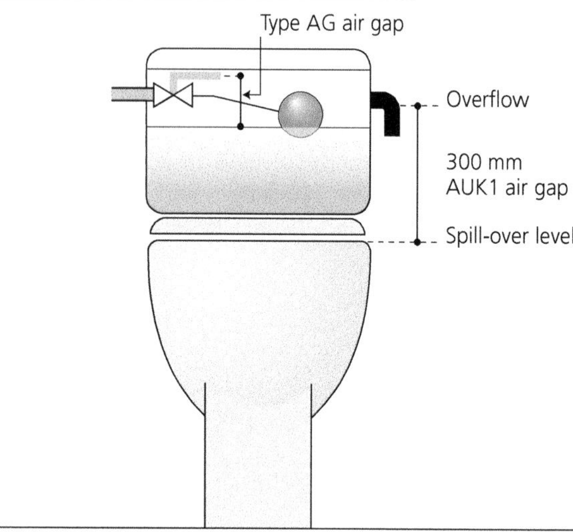

AG and AUK1 air gaps

- AUK2
  - The air gap between the outlet of a basin, bath or bidet tap and the spill over level.
  - Protects Category 1 against Category 3.

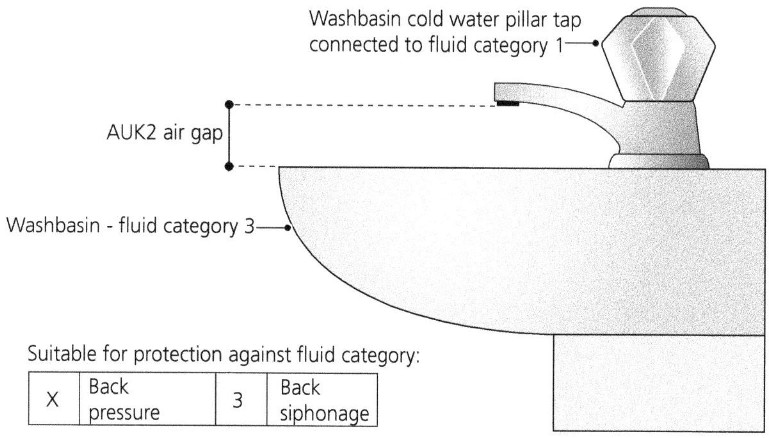

Suitable for protection against fluid category:

X	Back pressure	3	Back siphonage

AUK2 air gap

Exam tip

A good way to remember AUK2: (second letter of the alphabet) basin, bath and bidet.

- AUK3
  - The air gap between the outlet of the kitchen tap and the spill over level of the sink.
  - Protects Category 1 against Category 5.

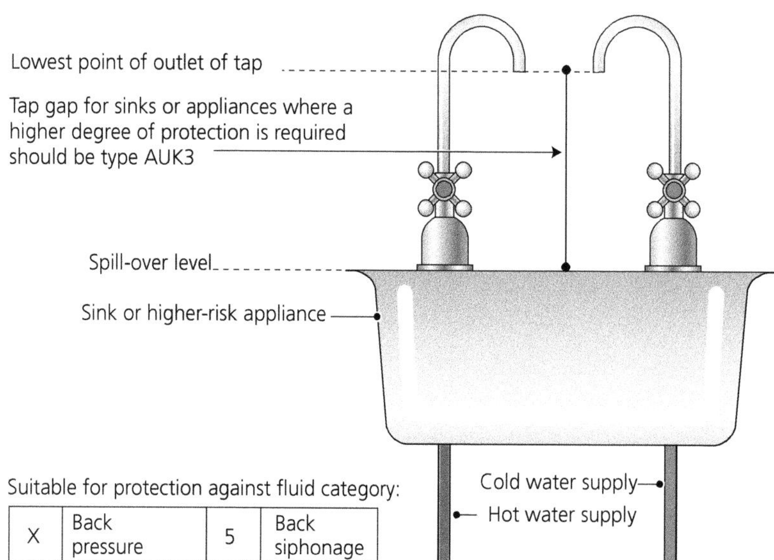

Suitable for protection against fluid category:

X	Back pressure	5	Back siphonage

AUK3 air gap

Now test yourself TESTED

You are installing a new outside tap for a customer.

3 What category of water is the hazard?
4 What is used to protect the Category 1 water?

LO3 Install cold water systems and components

Topic 3.1 Prepare for the installation of systems and components; Topic 3.2 Install and test systems and components

REVISED

There are parts of this Learning Outcome, mainly in Topic 3.1 and 3.2, that will be carried out as workshop activities for you to complete. You may still be questioned on parts of them in the exam, so do try to understand the concepts explained and undertaken in these activities. To back up the workshop activities, some key information is outlined below.

Table 3.7 Symbols found on drawings

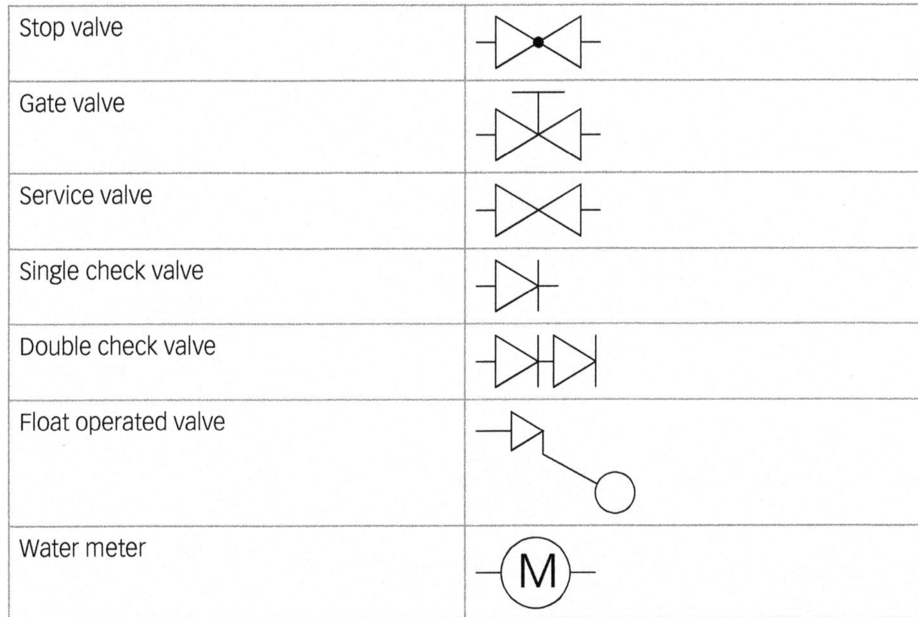

Stop valve	
Gate valve	
Service valve	
Single check valve	
Double check valve	
Float operated valve	
Water meter	M

Isolation valves

Service valves (spherical plug valves):

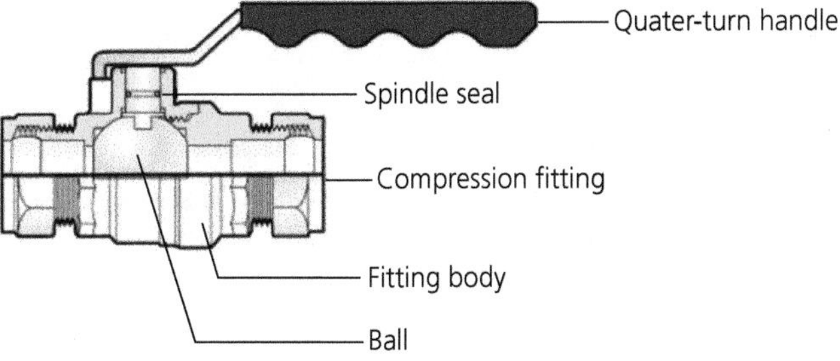

Service valve

- The Water Regulations state that a service valve should be installed prior to every float operated valve.
- They are also installed prior to appliances and taps to allow maintenance.
- Quarter turn by handle or screwdriver.

BS 1010 Stop valve:

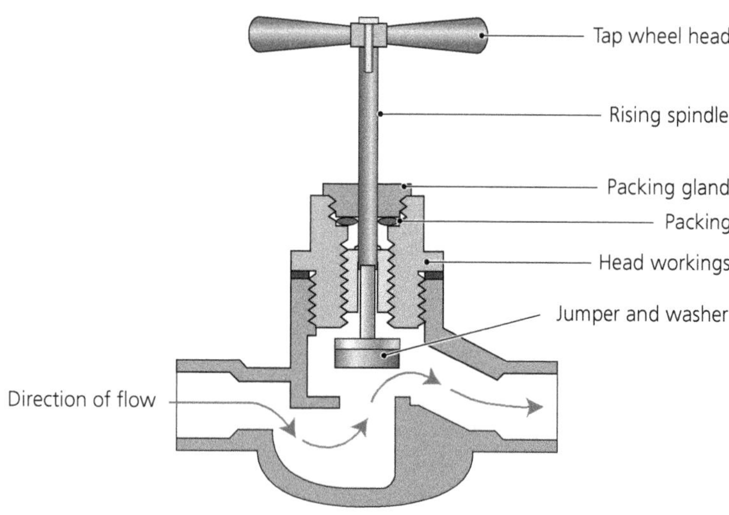

Stop valve

- High pressure isolation valve.
- Brass body, rising spindle with packing glands.
- Installed on incoming mains (rising mains).

BS 5154 Full-way gate valve:

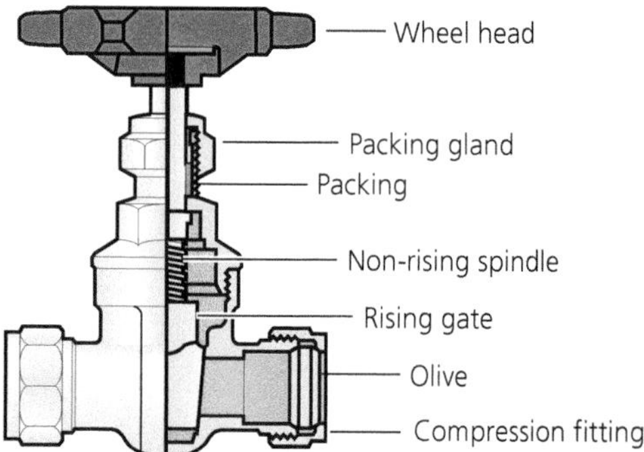

Gate valve

- Low pressure isolation valve.
- Full bore flow.
- Installed on the outlets of a cold water storage cistern (cold feed and cold distribution).

Check your understanding

9 What must be installed prior to every float operated valve?

Float operated valve:

- Used to control the flow of water into a cistern to a predetermined water level.
- Bottom or side entry.

BS 1212 Part 1:

- Float operated valve.
- Not used in domestic cisterns because the air gap cannot be guaranteed as the water outlet is under the body.
- Both use a piston.

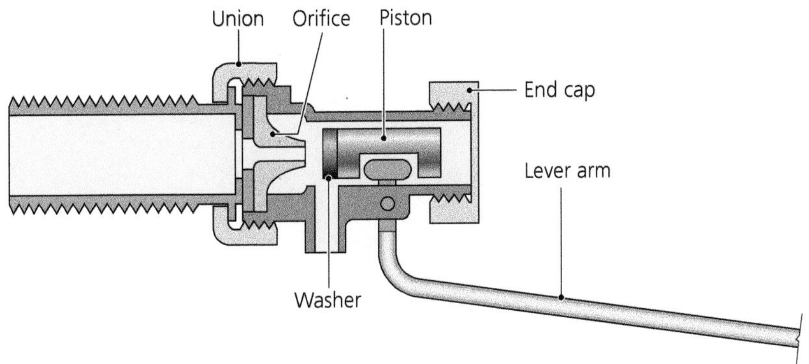

Portsmouth style

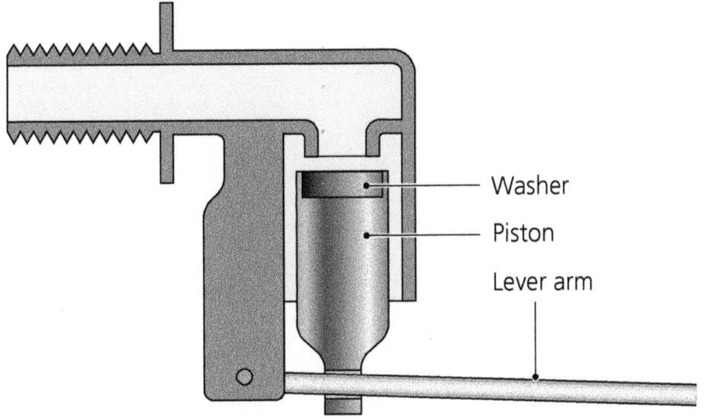

Croydon style

BS 1212 Part 2:

- Float operated valve.
- Water outlet from the top to allow for the required air gap.
- Use a diaphragm.
- Different orifice for high pressure (white) and low pressure (red).

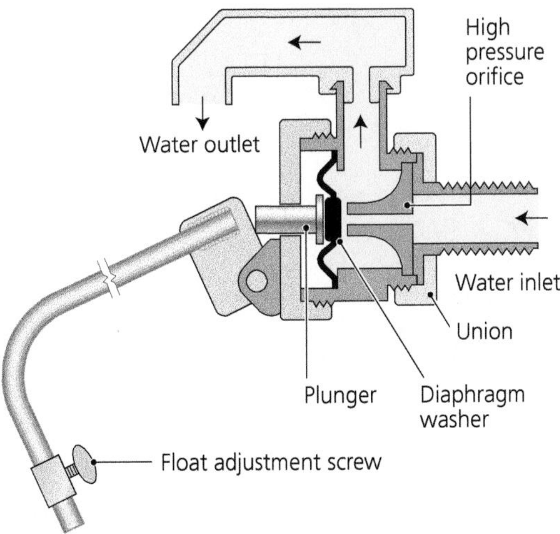

BS 1212 Part 2 float operated valve

BS 1212 Part 3:

- Float operated valve.
- As Part 2 but made of plastic.
- Different orifice for high pressure (white) and low pressure (red).

BS 1212 Part 4:

- Float operated valve.
- Torbeck equilibrium valve.
- Equal pressure either side of the diaphragm.
- Used to stop water hammer in a high-pressure system.

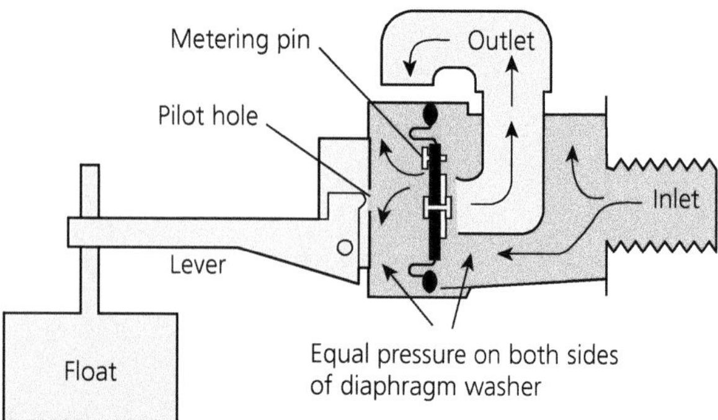

BS 1212 Part 4

Now test yourself TESTED

5 How do you make the adjustment between a high pressure (main) installation and a low pressure (cistern) fed installation when installing a BS 1212 Part 2 float operated valve?

Taps

(Hot = left-hand side, cold = right-hand side)

Table 3.8 Taps

Pillar	+ Thread under tap body + Basin, bath and bidet + Secured by back nut
Bib	+ Thread to rear of body + Back plate elbow required to mount tap to wall + Used to supply cleaner's sinks
Outdoor tap	+ Also known as a 'hose union bib' tap or 'outside' tap + Requires a double check valve to protect the Category 1 water + Requires insulating as the tap and pipework are vulnerable to freezing
Mixer	+ Water mixes inside the tap body + Risk of contamination Category 2 to Category 1 + Equal pressure from hot and cold supplies
Bi-flow mixer	+ Water mixes on exit + Two individual tubes inside spout + Common at kitchen sink + Unequal pressures are okay with this tap (mains cold, low pressure hot)

Table 3.9 Tap workings

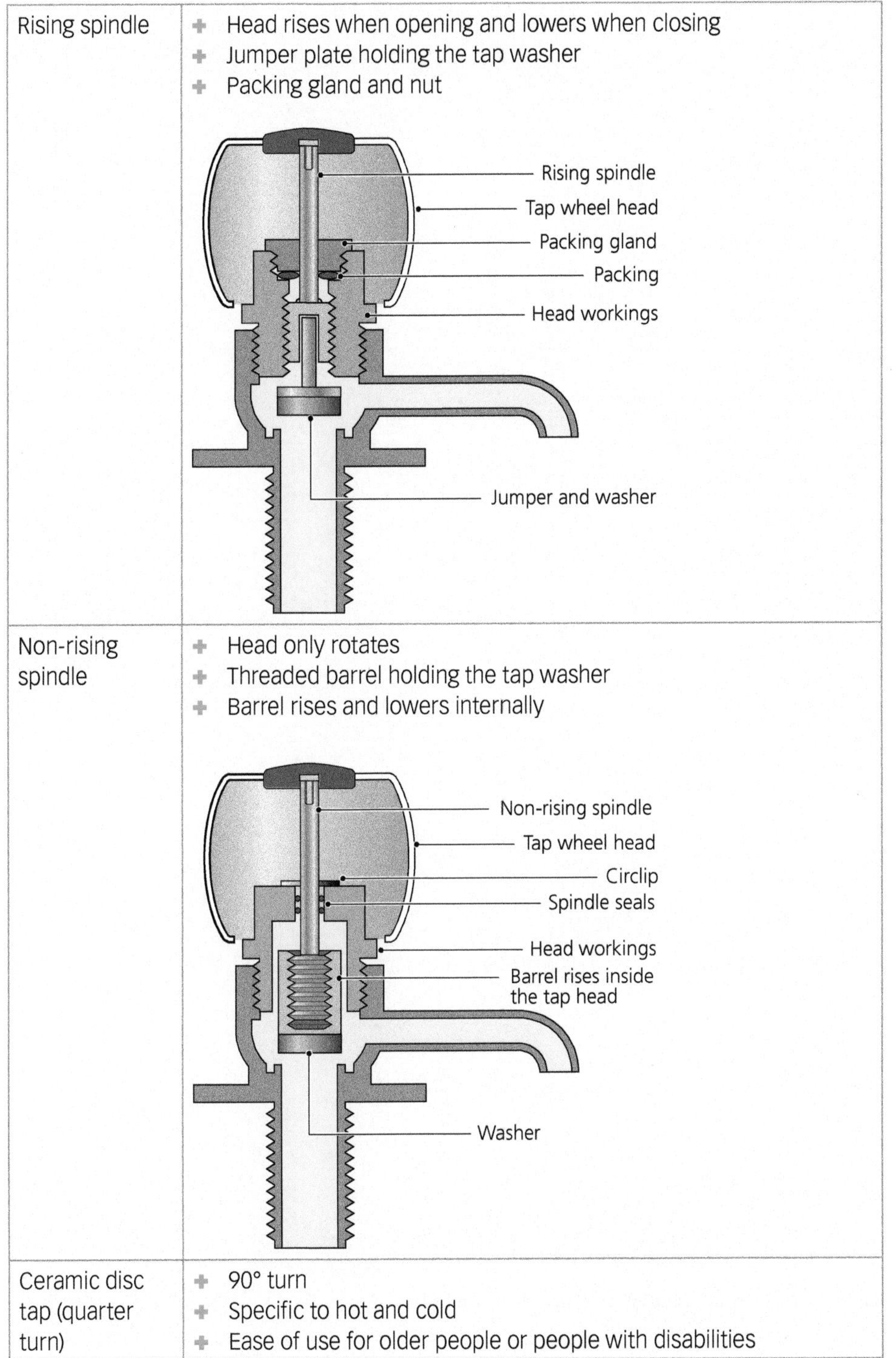

Rising spindle	+ Head rises when opening and lowers when closing + Jumper plate holding the tap washer + Packing gland and nut
Non-rising spindle	+ Head only rotates + Threaded barrel holding the tap washer + Barrel rises and lowers internally
Ceramic disc tap (quarter turn)	+ 90° turn + Specific to hot and cold + Ease of use for older people or people with disabilities

Other components

Table 3.10 Other components

Drain valve	+ The Water Regulations require a drain off to be installed at every low point in a system + Soldered, compression, push fit or threaded Jumper and washer Packing gland Rising spindle
Water softener	+ The only true way to remove water hardness (calcium carbonate) + Installed on the rising main + Ion exchange (resin beads) + Calcium carbonate removed and then back washed out overnight + Kitchen sink must be left unsoftened

Water conditioners	+ Do not remove the hardness but change the chemistry make up by suppressing the limescale
	+ Types of conditioners:
	  + magnetic
	  + electrolytic
	  + electronic
	  + electrochemical

Water filters	+ Installed to improve the taste, colour or odour of the water; jug filters are the basic style
	+ Plumbed in filters:
	  + activated carbon
	  + ion exchange
	  + sediment
	  + reverse osmosis
	  + distillation
	  + disinfection
	+ Each designed to remove a specific element

Water meter	+ Installed at the property boundary or, in some cases like flats, may be installed internally

Now test yourself

6 What problems might a customer encounter if they do not treat the hard water in their system?

Insulation

Frost protection is an insulation requirement of system components. It merely delays the freezing process.

+ Insulate to protect from frost in winter and prevent undue warming in summer
+ The thicker the insulation, the longer the heat energy is contained.
+ All cold water pipework outside the thermal envelope must be insulated.

The table below gives the recommended thicknesses of insulation to be used. For example, a pipe with 15 mm external diameter would need 20 mm of insulation with a thermal conductivity of 0.020, 30 mm of insulation with a thermal conductivity of 0.025 and so on.

Table 3.11 Recommended insulation thicknesses

External diameter of pipe in mm	Thermal conductivity of insulation material at 0°C in W/m²K				
	0.020	**0.025**	**0.030**	**0.035**	**0.040**
15	20 (20)	30 (30)	25* (45)	25* (70)	32* (91)
22	15 (9)	15 (12)	19 (15)	19 (19)	25 (24)

*15mm pipe limited to 50% ice formation after 9, 8 and 7 hours respectively.
Figures in brackets show the minimum thickness for 12 hour frost protection.

Check your understanding

10 List the places that are outside the thermal envelope of the building.

Inspection and testing

Table 3.12 Inspection and testing

Visual inspection	+ All pipework has been connected + No open ends + Sufficient clipping for pipework + Check with drawing or plans + No evident leaks or drips
Soundness test	**The Water Regulations and BS EN 806** Copper and LCS pipework: + Fill system with water + Leave for 30 mins (temperature stabilisation) + Connect hydraulic bucket + Pump pressure up to 1.5 times the working pressure + Leave for one hour + No pressure drop = success Plastic (polybutylene) pipework Test A: + Fill system and pressurise to one bar + Check for leaks + After 45 mins increase pressure to 1.5 times operating pressure + Leave for 15 mins + Release pressure to 1/3 and leave for 45 mins + No final pressure drop = success Plastic (polybutylene) pipework Test B: + Fill system and pressurise to working pressure + Leave for 30 mins and check pressure + Leave for a further 30 mins and check pressure + If the pressure loss is less than 0.6 bar with no visible leaks = success Any pressure drop or leak requires rectification and retesting

Commissioning	+ Once the system has been tested for soundness and passed:
	  + flush through with cold water – remove debris and flux
	  + refill system
	  + inspect water levels in cisterns
	  + inspect valves and ensure they are holding (FOV, taps and drain offs)
	  + check flow rates at outlets (flow or weir cup – manufacturer's instructions)
	  + check pressure at outlets
	+ Pressure test would be carried out using a pressure gauge to confirm the outlet pressure against the manufacturer's instructions
	+ Manufacturer's instructions are always consulted when commissioning, replacing parts or fault finding
	+ Verbal instructions are also useful from team members and the manufacturers

Check your understanding

11 Why is plastic pressure pipe tested for soundness at a lower pressure than rigid pipework?

Worked example

If the working pressure of a system is 4 bar, what pressure should a system be tested at if the pipework is copper?

1.5 times the working pressure = 4 bar × 1.5 = 6 bar test pressure

Now test yourself

TESTED

7 If the working pressure is 3 bar, what pressure should a system be tested at if the pipework is copper?

Exam-style questions

1 What is likely to occur if a tap spout is below the spillover level of the basin?

a Water temperature could rise
b Flow rate would decrease
c Backflow could occur
d Water pressure would decrease

2 What are the two MAIN considerations for the incoming mains cold water supply?

a Flow and colour
b Temperature and pressure
c Taste and temperature
d Pressure and flow

3 Which British Standard offers guidance on the capacity of cold water storage cisterns?

a BS 8000 c BS EN 12056
b BS EN 806 d BS 1212

4 What category of water is wholesome, potable drinking water?

a 1 b 2 c 3 d 4

5 What is the name of the cold water main pipe that connects the mains supply in the road up to the boundary meter?

a Service pipe
b Rising main
c Communication pipe
d Supply pipe

6 What category of water presents a severe health risk?

a 2 b 3 c 4 d 5

7 What mechanical backflow protection would be used to protect against Category 3 water?

a RPZ valve
b Gate vale
c Single check valve
d Double check valve

8 Which one of the following is not a factor in the Water Supply (Water Fittings) Regulations?

a Contamination of water
b Misuse of water
c Pressure of water
d Waste of water

9 According to the Water Supply (Water Fittings) Regulations, what must be installed prior to every float operated valve?

a Gate valve c Drain off valve
b Stop valve d Service valve

10 You are testing a customer's cold water system and you trace the pipework back which is connected to the incoming main. What type of system is this?

a Indirect system
b Direct system
c Low pressure system
d High cost system

11 What type of system can a full way gate valve be installed in?

a Low pressure c Direct system
b High pressure d Interrupted system

12 Why can you not install a BS 1212 Part 1 float operated valve in a domestic cistern?

a Does not allow high pressure water through it
b Made of the wrong material
c Cannot maintain the required air gap
d Flow rate is too poor

13 Where would you install a stop valve in a domestic property?

a The cold distribution c The rising main
b The cold feed d Prior to a bath tap

14 When linking cold water storage cisterns together, which of the following installations will NOT cause stagnation?

a

b

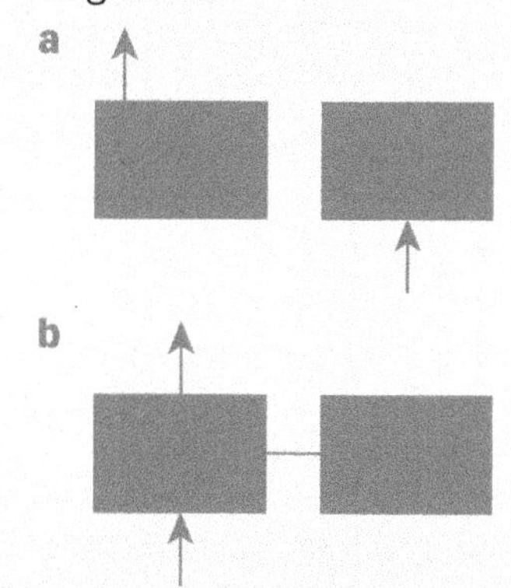

c

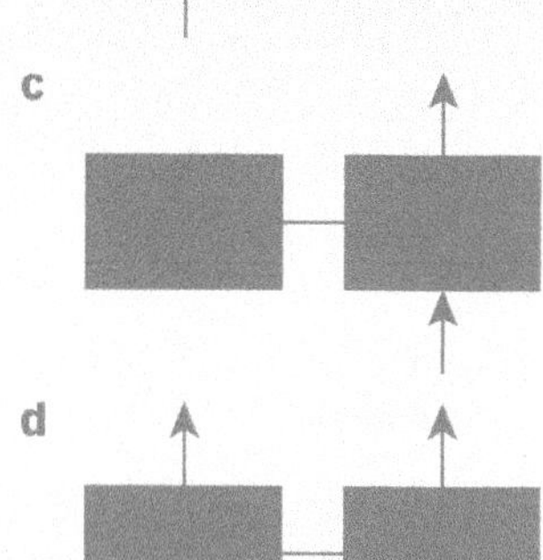

d

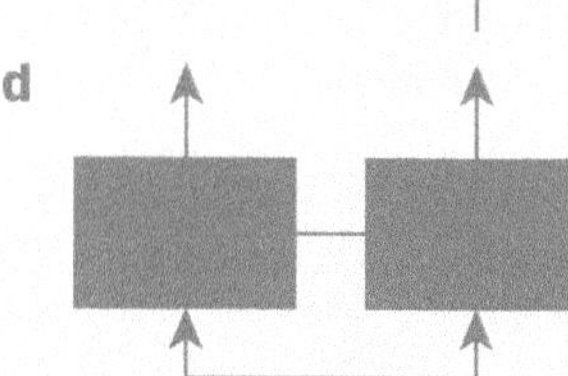

15 What is the correct depth that the incoming cold water mains must be laid between?

a 750 – 1350 mm
b 650 – 1250 mm
c 550 – 1150 mm
d 450 – 1050 mm

16 What air gap is being identified on this basin?

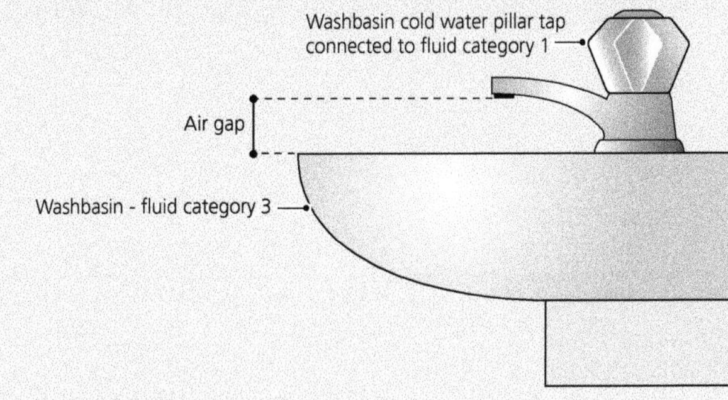

a AUK1 **b** AUK2 **c** AUK3 **d** AG

17 The working pressure on a system is 2 bar. This pipework has all been installed in copper. What pressure does the system need to be pressurised to for a soundness test?

a 1 bar **b** 2 bar **c** 3 bar **d** 4 bar

18 What is the only true way to remove the hardness from the water within a property?

a Install a water filter
b Install a water conditioner
c Install a water softener
d Install a water retarder

19 When commissioning a new cold water system in a customer's property, the system will need to have the pressure tested, the flow rate recorded and water levels set. Which of the following should also take place as part of the commissioning process?

a The pipework will need to be clipped
b The pipework will need to be soldered
c The pipework will need to be flushed
d The pipework will need to be painted

20 Which design of tap will need protection against Category 2 water contaminating Category 1 water?

a Ceramic disc
b Bib
c Mixer
d Pillar

4 Hot water (Unit 215)

You need to familiarise yourself with the sources of information relating to domestic hot water systems, appliances and controls. You will also need to understand the testing, decommissioning and maintaining of hot water systems to keep them in good condition. Ask yourself:

+ What is a safe temperature for hot water and how is it controlled?
+ How do I select a hot water system?
+ How do direct and indirect systems work?
+ Why is maintenance important?

LO1 Understand hot water systems and their layouts

Topic 1.1 Sources of information relating to work on hot water systems

REVISED

Statutory regulations

Building Regulations Approved Document G3:

+ G3 Part 1 – Hot water must be supplied to a basin in, or adjacent to, a WC and food preparation area.
+ G3 Part 2 – Hot water systems must be able to resist the effect of temperature or pressure during normal use.
+ G3 Part 3 – Hot water temperature must not reach 100°C. Any discharge must be visible and safely conveyed.
+ G3 Part 4 – Water temperature to a bath must not exceed 48°C (new build and conversions).

Building Regulations Approved Document L1A and B:

+ Conservation of fuel and power to reduce CO_2 emissions.
+ Insulation to reduce heat loss (hot water cylinder, pipework and building fabric).
+ System must be controlled (programmers, thermostats, zone valves and so on).

The Water Supply (Water Fittings) Regulations:

This reflects the two previous regulations, as well as:

+ unvented hot water systems
+ open vents (minimum diameter 19 mm, vent over cold water cistern)
+ distribution temperatures (stored at 60°C and distributed at 55°C)
+ at any outlet, 50°C should be reached within 30 seconds of opening
+ hot water pipework should not form a 'dead leg' (15 mm pipe cannot be longer than 12 m and 22 mm pipe cannot be longer than 8 m)
+ 4% expansion must be accommodated (cistern or vessel).

The Gas Safety (Installation and Use) Regulations:

+ Safe installation, testing and maintenance of gas appliances.
+ They must be 'Gas Safe' registered, which is a legal requirement.

The IET (18th Edition) Wiring Regulations:

+ Safe installation, testing and maintenance of wiring and electrical appliances.
+ Installed according to BS 7671.
+ Controls, immersion, spurs.

Industry standards

BS EN 806 parts 1 to 5:

- Design, installation, testing and maintenance of hot water systems.

BS 8558:

- Complements BS EN 806 in the fact that it also covers the design, installation, testing and maintenance of hot water systems.

The Domestic Heating Compliance Guide:

- Guidance to help with Building Regulations Part L.
- Required controls (boiler thermostat, cylinder thermostat, programmer, circulator, zone valves).

Manufacturers' technical instructions

- These are key documents when installing, testing and maintaining appliances.
- Unvented hot water systems **must** be fitted, commissioned and maintained as per manufacturers' technical instructions.
  - Minimum pressure and flow rates.
  - Minimum size of incoming cold supply.
  - Minimum size of hot distribution.
  - Required heat input and heat recovery time.
  - Electrical requirements.
  - Operational controls.
  - Calculations for discharge pipework.
  - Fault finding.
- They are usually available online.

Exam tip

Heat recovery time means the time to heat the water from cold. This is linked to Building Regulations Part L. If a question asks about the heat recovery time for a hot water cylinder, it should be 20–25 mins.

Check your understanding

1 Which category of water is hot water according to the Water Regulations?

Now test yourself TESTED

1 In an appliance where hot water and cold water mix, there is a possibility of the hot water entering the cold water system. What does the Water Regulations classify this cross-flow as?

Topic 1.2 Hot water systems and components

REVISED

There is some crossover between this unit and Chapter 3, Cold water (Unit 214). Refer back to this chapter to see more details on valves, taps and backflow prevention.

Factors to consider when selecting hot water systems

Table 4.1 Factors to consider when selecting hot water systems

Occupancy	+ The number of people living in the property affects the amount of water used + It affects the cold water storage cistern size, hot water cylinder size and the supply/distribution pipework size + The more people, the more appliances
Type and size of building	+ Position of the boiler and hot water cylinder or source of hot water + 'Dead leg' lengths – need for secondary circulation + Surface or hidden pipework + Number of bathrooms, WC and other outlets + Extension or existing building

Available services	What is the existing system Replace or extend existing system Increase pipe size or storage Electrical supply in location Gas, oil, LPG, incoming main, hot distribution, cold feed sizes and location
Typical sizes	Open vent – minimum size 19 mm in diameter Cold feed – minimum size 22 mm (must equal or be bigger than the hot distribution) Hot distribution – minimum size 22 mm Cold mains for unvented system – 22 mm Hot feed to taps normally 15 mm (except bath, which is 22 mm) Hot feed to showers normally 22 mm to within 1 m, then reduces to 15 mm Most common hot water cylinder: 450 × 900 mm

Check your understanding

2 Why is it so import to follow manufacturers' instructions carefully

3 When installing a hot water system, one of the considerations that has to be made is occupancy. How will the occupancy affect the choice of hot water cylinder and cold water storage cistern?

Exam tip

The chart below, outlining centralised and localised hot water systems, is particularly important to be able to understand, as it shows the choice and styles of systems that are available at the point of design.

Types of systems

This diagram shows the different types of centralised and localised hot water system that may be installed.

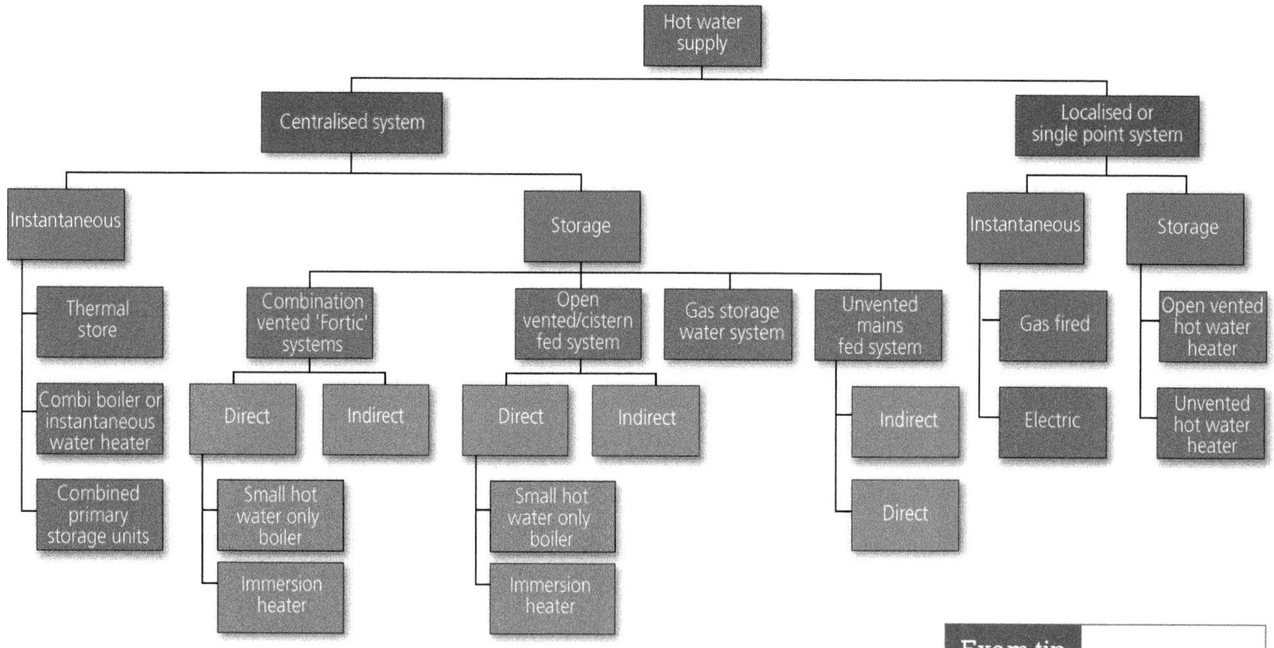

Hot water systems

Check your understanding

4 What is the primary function of an open vent?

5 What type of system is likely to be installed in a smaller property that only has one bathroom and therefore less demand?

Exam tip

You must be able to relate the key terms to the different systems. For example, 'centralised' means a central point within a property delivering hot water to various outlets.

Now test yourself TESTED

2 A customer has been up into their loft where the cold water storage cistern is located. They notice a 22 mm pipe coming up into the loft, over the cistern and going through the cistern lid. They have asked you what pipe it is and what function it has. Can you explain?

Direct hot water systems

Boiler heated direct system:

- No coil.
- The primary water is also the secondary hot water.
- Hot water could be 'dirty' as it has gone through the boiler.
- Low pressure system.
- Old system which has been replaced by immersion heated direct systems.

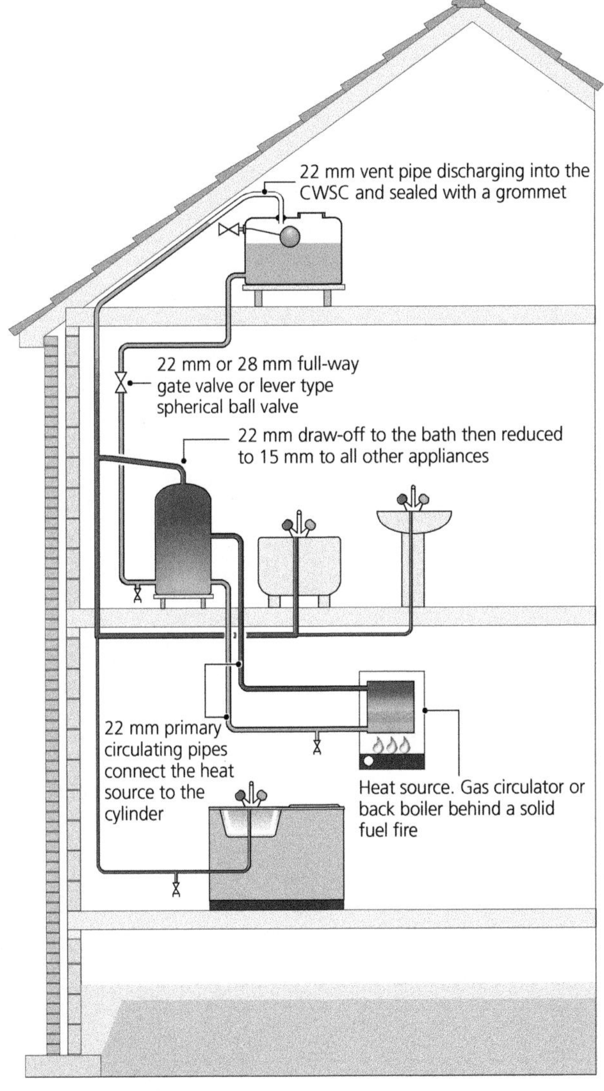

Direct boiler heated system

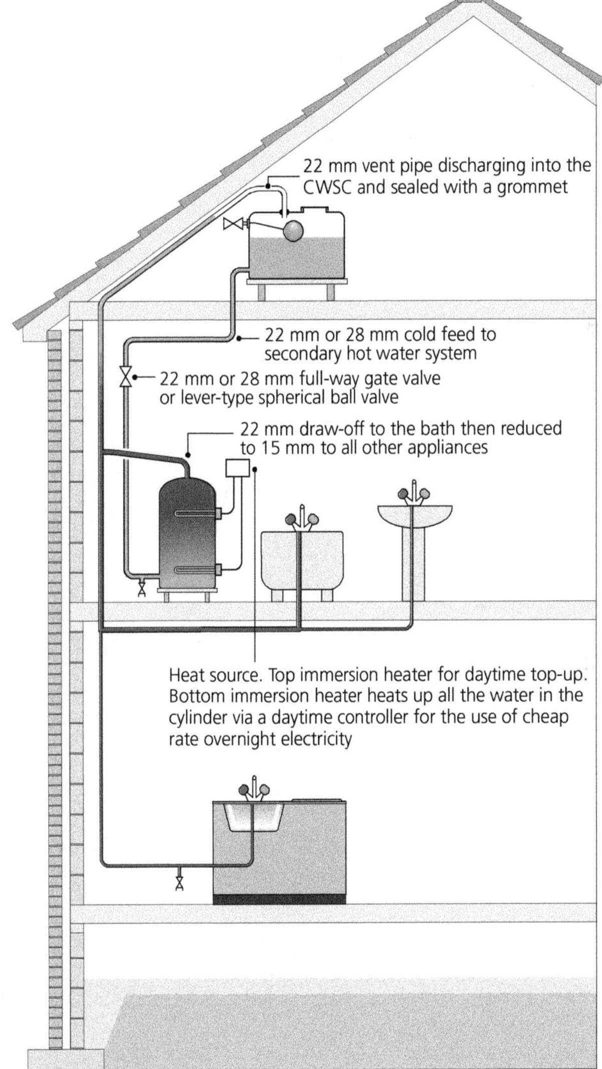

Direct immersion heated system

Immersion heated direct system:

- Modern direct system used where there is no boiler available.
- Low pressure system.
- Two immersions allow a back-up heat source and also flexibility to only heat up part of the cylinder.

Typical mistake

Remember that 'direct system' means the water is heated **directly** by the heat source (boiler or immersion).

Check your understanding

6 Is a direct hot water system, heated by an immersion heater, a high or low pressure system?

Indirect hot water systems

Open vented **double feed system:**

- Double feed.
- First feed from the cold water storage cistern to the base of the hot water cylinder – cold feed.
- Second feed from feed and expansion cistern to the primary circuit of the boiler – cold feed.
- Modern double feed systems have a pump on the primary pipework – older systems relied on gravity circulation.
- Cylinders and systems must comply with Building Regulations Approved Document L 1A and B.
- The boiler heats the primary water, which heats the coil, which in turn heats the hot water.

Open vented A system open to atmospheric pressure by the use of an open vent pipe.

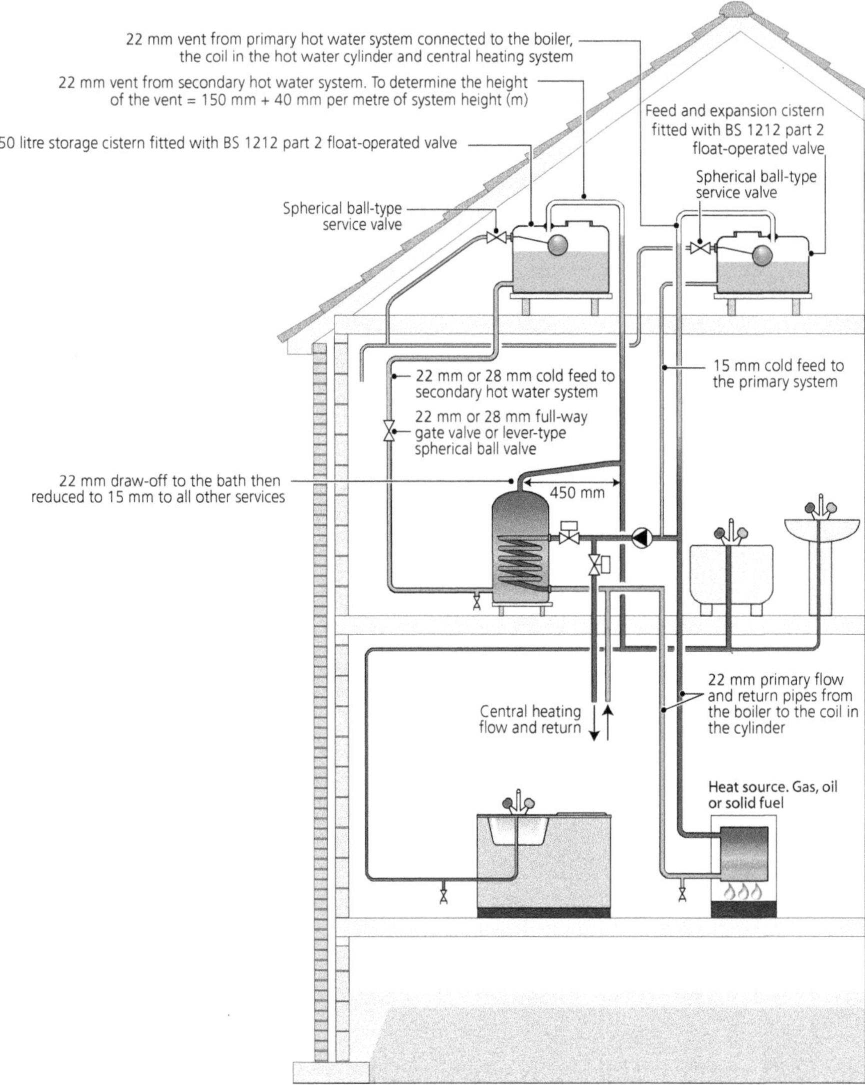

Double feed system

Open vented single feed system:

- Also known as a primatic system.
- Single feed from the cold water storage cistern to the base of the hot water cylinder – cold feed supplies both the primary and secondary water.
- Primary and secondary water are separated by an air bubble lock.
- Old system, no longer used.
- The boiler heats the primary water, which heats the heat exchanger vessel, which in turn heats the hot water.
- If the air bubble is lost due to excessive expansion, then the waters mix and become dirty.
- Different orifice for high pressure (white) and low pressure (red).

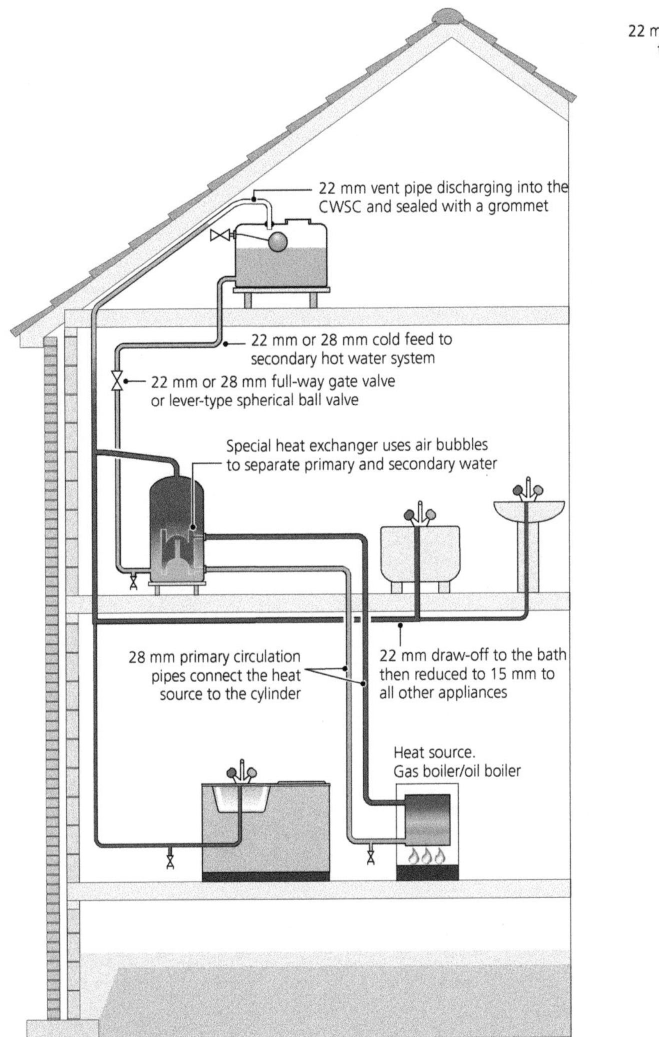

Single feed system

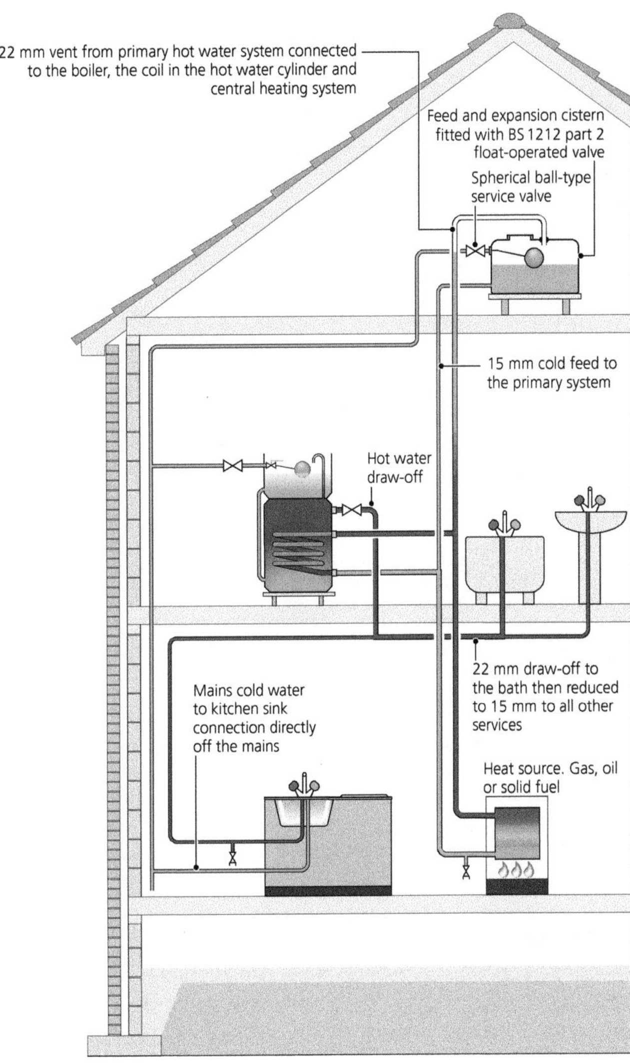

Combination unit

Open vented combination unit:

- An all-in-one unit.
- Known for very low pressure as cold water cistern is located on top of the cylinder, giving no head height.
- Used in flats and holiday cabins.
- The boiler heats the primary water, which heats the coil, which in turn heats the hot water.
- Also known as 'fortic cylinders'.

Typical mistake

Be careful not to mix up a combination boiler with a combination unit!

Thermal store (water jacket):

- Instantaneous water heater.
- Works in the opposite way to an indirect cylinder.
- The boiler heats the primary water which is now on the outside of the coil and fills the cylinder.
- This heats the coil, which in turn heats the cold water flowing through the coil or heat exchanger.

Stored Water heated and kept prior to demand (hot water cylinder).

Instantaneous Water heated on demand to the outlets (combination boiler or thermal store).

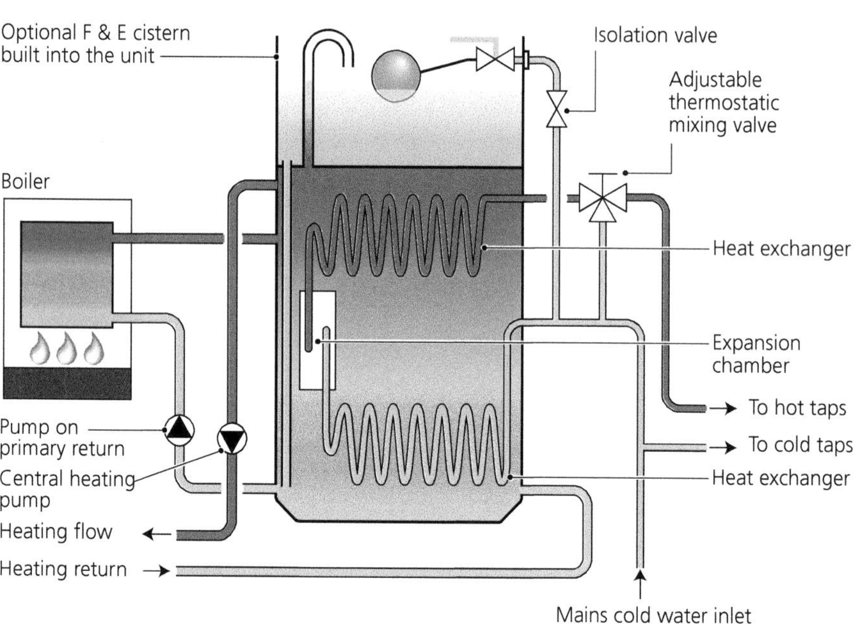

Thermal store

Check your understanding

7 Name the two cisterns installed in a double feed system.

Now test yourself TESTED

3 You go to a customer's property where an old primatic cylinder is installed. The customer would like to upgrade their system to a modern low pressure system using their existing boiler. What type of cylinder would you install?

Exam tip

'Indirect system' means the water is heated **indirectly** by the heat source. The boiler heats the primary water, the primary water heats the coil (cylinder heat exchanger), the coil heats the water. Knowing this will help you gain marks from questions testing your understanding of hot water systems.

Exam tip

Many students struggle to identify systems and their component parts. If you get a system to identify, follow the pipework and flow of water through the system with your finger to help you identify it.

Instantaneous hot water heaters

Single point water heater:

- Localised system.
- 8–11 kW heaters.
- Serves one appliance only.
- A small amount of heated water is kept.

Localised Hot water heated and delivered at the point of use to the outlet (single point water heater).

- The spout acts as the open vent.
- On the under the sink heater version, a special tap needs to be installed to act as the open vent.
- Drips from the spout as water expands.
- Inlet controlled – the tap allows water into the unit, which pushes the hot water out of the spout.

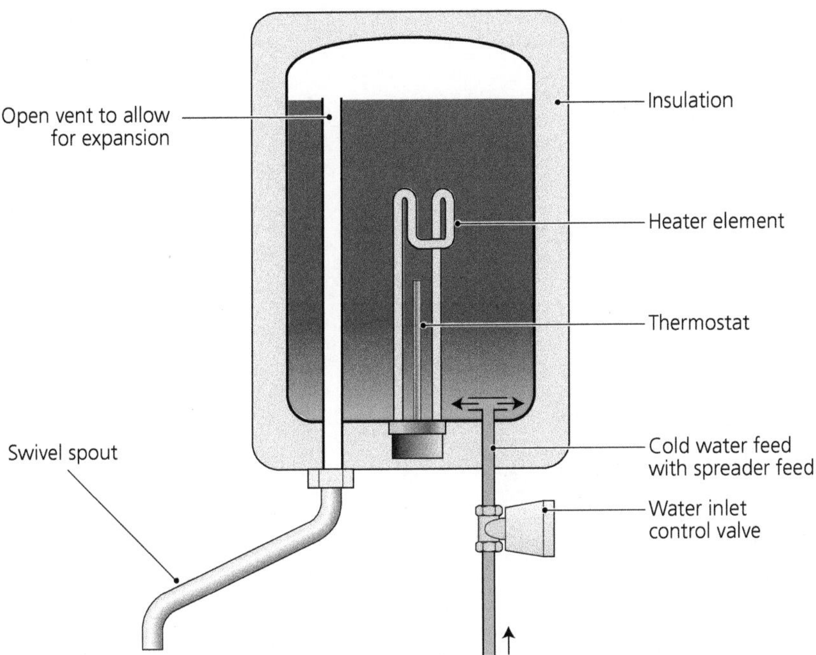

Single point heater

Multi-point water heater:

- Localised system.
- Serves several appliances locally.
- Outlet controlled – the tap is opened, allowing the hot water to flow.

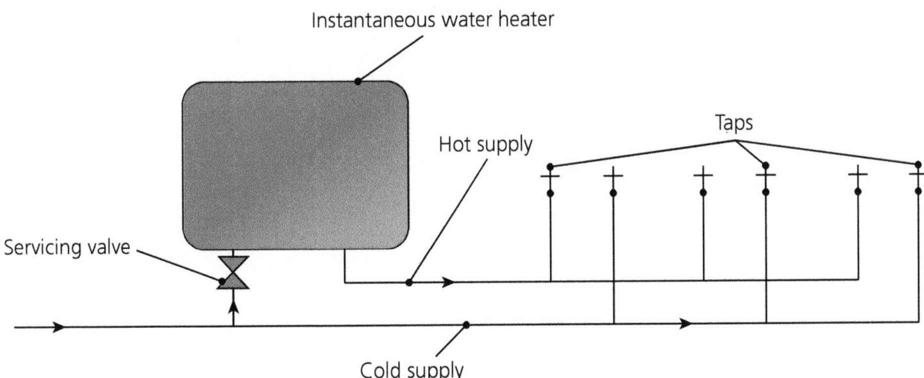

Multi-point heater

Combination boiler:

+ Centralised system.
+ Hot water heated on demand via plate heat exchanger.
+ Hot water priority – the boiler can heat the central heating or hot water but not at the same time.

Centralised Hot water delivered from a central point to the outlets (cylinder or combination boiler).

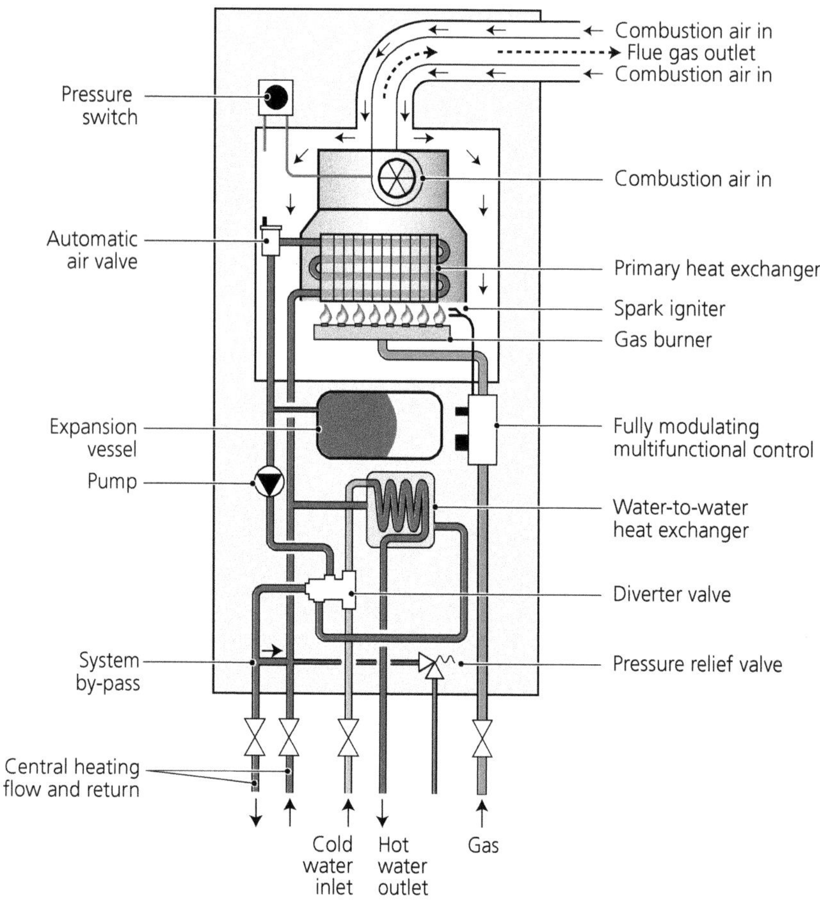

Combination boiler

Check your understanding

8 What system would you install in a small, two-bedroom property with one bathroom, where space was limited?

Now test yourself TESTED

4 You are asked to price up a hot water heater for a small community hall. It requires hot water to be delivered to a remote kitchen sink. What heater would you suggest?

5 If the head height for a system is 3.0 m, how high does the open vent need to rise above the water level in the cistern?

Showers

Table 4.2 Showers

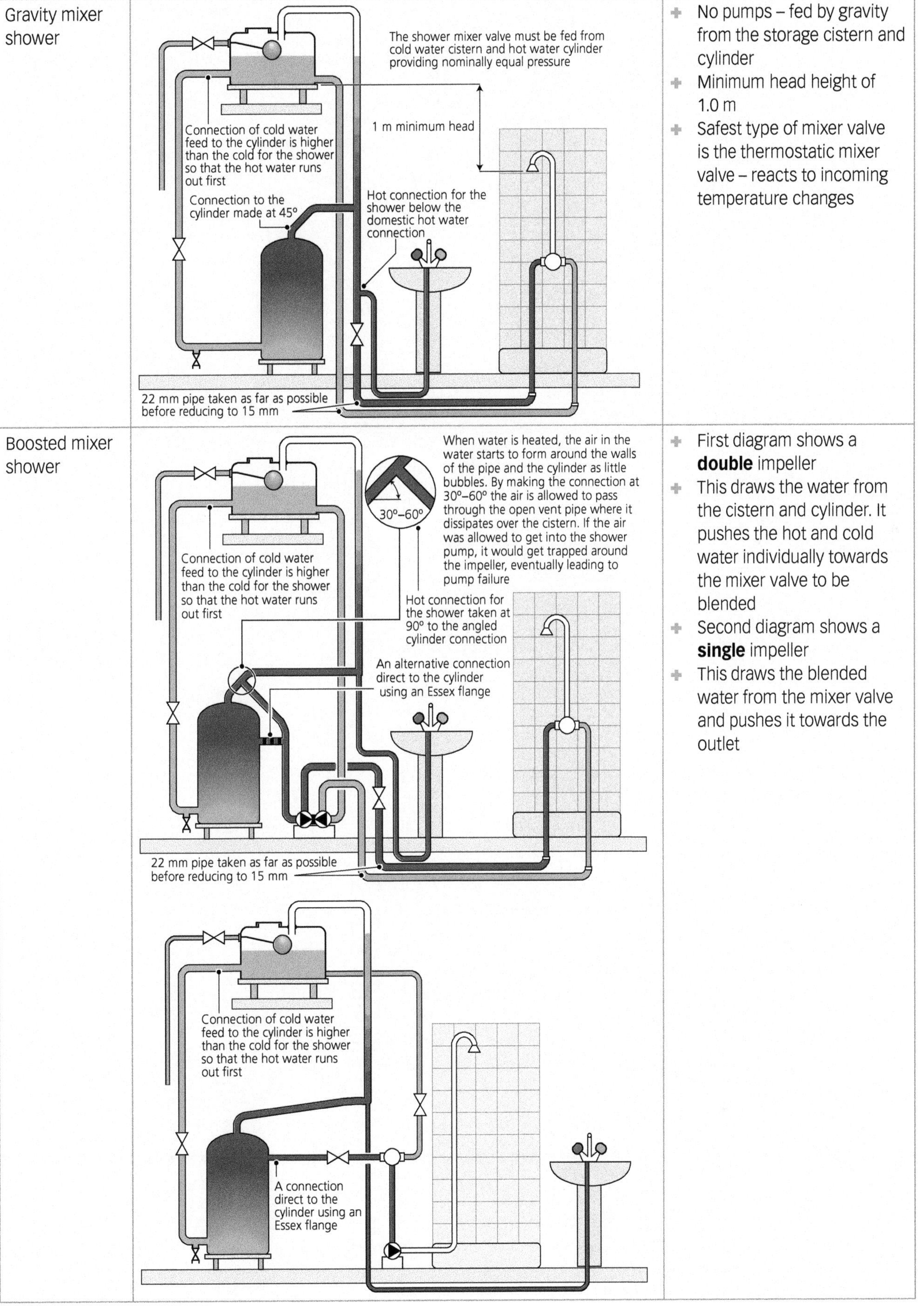

Gravity mixer shower		+ No pumps – fed by gravity from the storage cistern and cylinder + Minimum head height of 1.0 m + Safest type of mixer valve is the thermostatic mixer valve – reacts to incoming temperature changes
Boosted mixer shower		+ First diagram shows a **double** impeller + This draws the water from the cistern and cylinder. It pushes the hot and cold water individually towards the mixer valve to be blended + Second diagram shows a **single** impeller + This draws the blended water from the mixer valve and pushes it towards the outlet

<table>
<tr>
<td>Electric instantaneous</td>
<td>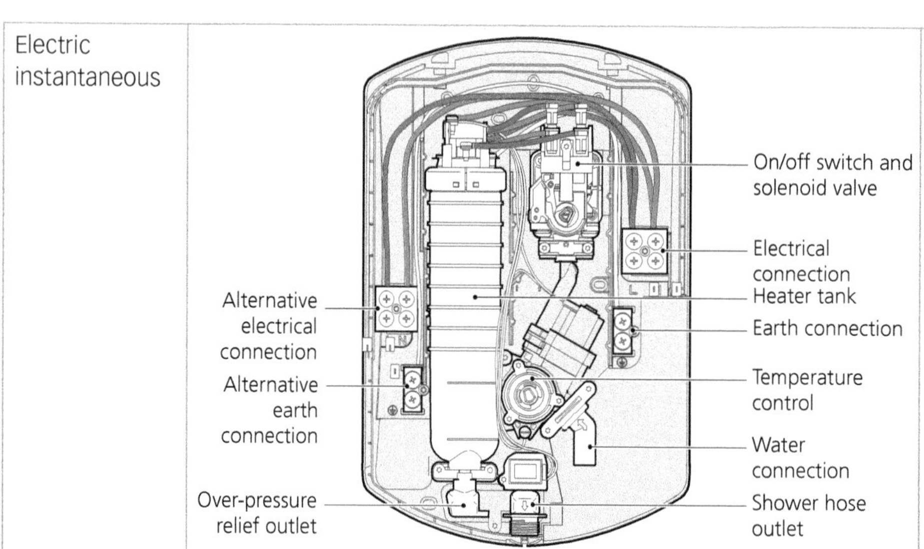
</td>
<td>
+ Supplied by cold water mains
+ Heats water up on demand
+ 8–11 kW heaters are common
+ Wiring needs to be installed by competent electrician
</td>
</tr>
</table>

Exam tip

Diagrams of shower installations sometimes come up in exams, so do note the single and double impellers on the boosted showers (the circle with a black triangle inside).

Check your understanding

9 A customer would like a shower installed in a new en suite bathroom, but the en suite only has a cold water mains supply. Instead of re-piping a hot water supply over to the room, what shower could be installed?

Topic 1.3 System safety and efficiency

REVISED

Temperature control

Boiler thermostat:

+ Located on the boiler.
+ Sets the temperature for the primary flow and return water.
+ Maximum temperature 85°C.

Cylinder thermostat:

+ Located 1/3 of the way up the cylinder.
+ Set at 60°C. If lower than 50°C, legionella is a risk. If higher than 60°C, limescale is a risk (it forms at 65°C+ in hard water areas).

Overheat thermostat:

+ Overheat thermostat (thermistor) is located on the return pipe in the boiler.
+ It puts the boiler into lock out if the boiler thermostat fails and the water temperature rises above 85°C.
+ An overheat thermostat on an unvented system cuts the power source to the system and closes to the zone valve if the cylinder thermostat fails. Normally at 85°C.

Pressure relief valve:

+ Located on a combi boiler and an unvented system.
+ Opens if the pressure in the system rises above the recommended pressure (around 3 bar).
+ Normally due to the failure of the expansion vessel.
+ Drips to outside on combi and into the tundish on an unvented system.

Unvented A sealed pressurised system with safety controls.

Tundish Part of the discharge pipework joining D1 and D2 together, offering a visual sight of any discharge.

Temperature pressure relief valve:

- Located on the top of an unvented cylinder.
- It empties the cylinder of hot water if the pressure or temperature get too high.
- Opens if the pressure relief valve fails (normally about 5 bar).
- Opens if the cylinder thermostat and the overheat thermostat fail (normally about 95°C).
- Pours into the tundish.

Three tiers of safety on an unvented system:

- Cylinder thermostat – 60°C.
- Overheat thermostat – 85°C.
- Pressure/temperature relief – 95°C.

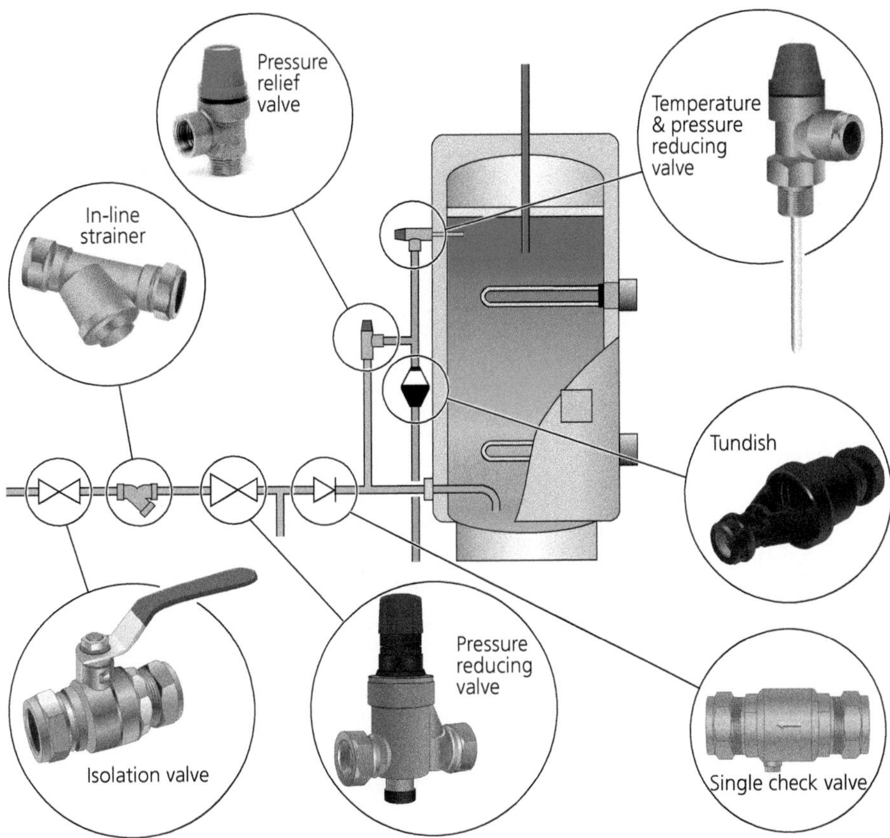

Unvented components

Thermostatic mixing valve – TMV or blending valve:

- Single valve installation most common.
- Maximum supply length after TMV is 2.0 m.

Temperatures:

- Bath 41 – 44°C.
- Shower – 43°C (domestic).
- Shower – 41°C (care homes).
- Basins – 43°C.
- Bidet – 38°C.
- Sink – 46–48°C.
- The sink needs to be hotter to kill bacteria and remove grease but leaves a risk of scalding.

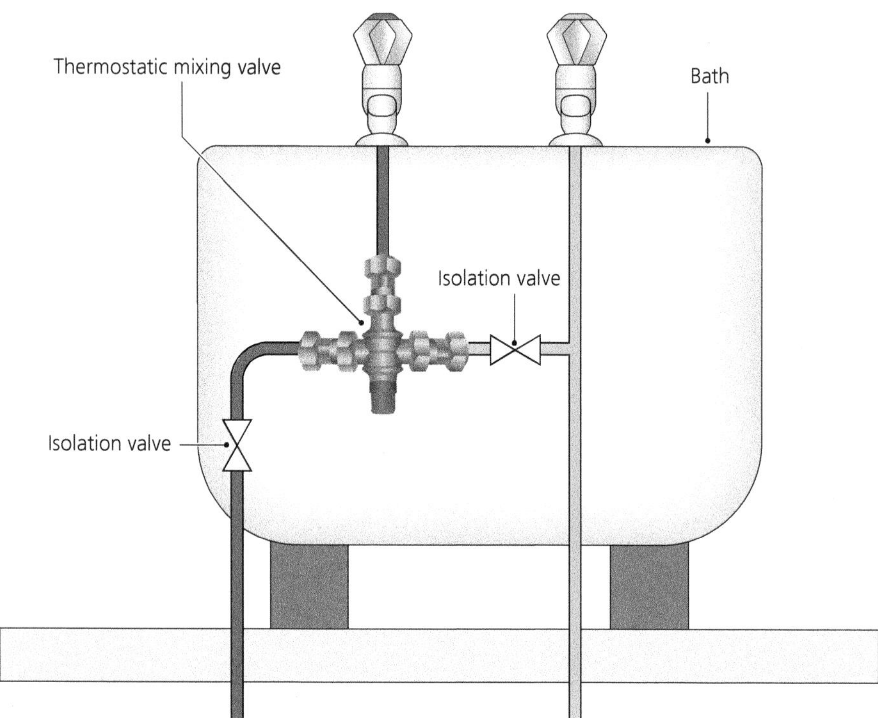

Single TMV

Group mixing installation:

- The operation of one appliance must not affect another.
- Care over legionella due to longer pipe runs.
- Regular disinfecting.

Temperatures:

- Group showers – 38–40°C.
- Group basins – 38–40°C.
- Neither must exceed 43°C.

Outlet temperature of appliances:

- These are set by the Building Regulations (Part G3).
- TMV2 is used domestically, while TMV3 is used for health care.

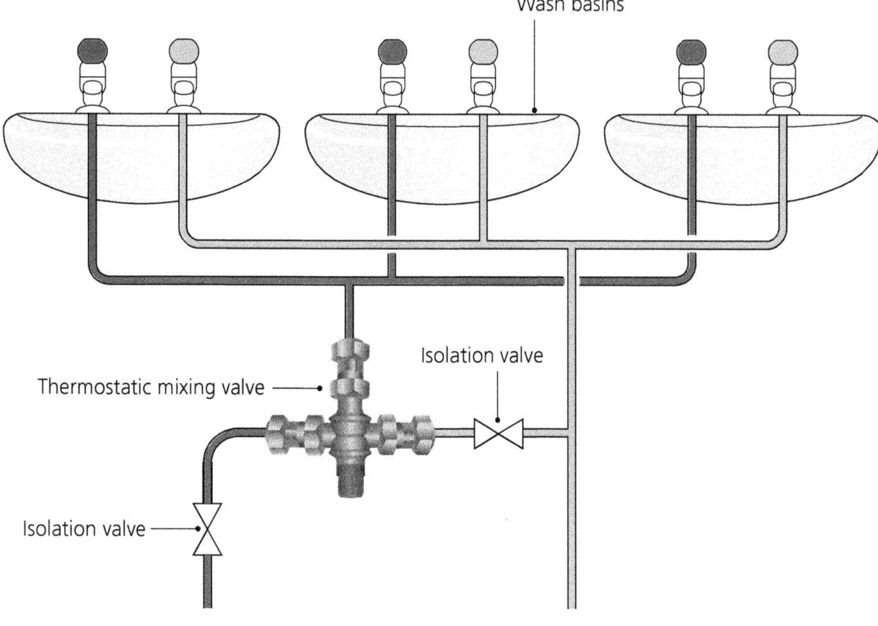

Group TMV

Installation requirements for pipework, cylinders, systems and cisterns are covered in:

- the Water Regulations
- BS EN 806
- BS 8558.

Check your understanding

10 Why is the cylinder stat set at 60°C and not 65°C?

11 When a new TMV is fitted before a domestic shower, what temperature should it be set at?

LO2 Install hot water systems and components

Topic 2.1 Prepare for the installation of systems and components; Topic 2.2 Decommission systems and components; Topic 2.3 Install and test systems and components; Topic 2.4 Replace defect components

REVISED

Learning Outcome 2 is largely practical and involves workshop activities. Use these practical tasks to learn important installation facts that will also help you answer some questions in the exam. You can refer back to LO3 in Chapter 3, Cold water (Unit 214), which covers components, inspection and testing of systems.

Here are some additional key points:

Table 4.3 Inspecting and testing

Visual inspection	**Check the following**: + all pipework has been connected + no open ends + sufficient clipping for pipework + check with drawing or plans + no evident leaks or drips + wiring correct and safe
Soundness testing	**The Water Regulations and BS EN 806** Copper and LCS pipework: + Fill system with water + Leave for 30 mins (temperature stabilisation) + Connect hydraulic bucket + Pump pressure up to 1.5 times the working pressure + Leave for one hour + No pressure drop = success Plastic (polybutylene) pipework Test A: + Fill system and pressurise to 1 bar + Check for leaks + After 45 mins increase pressure to 1.5 times operating pressure + Leave for 15 mins + Release pressure to 1/3 and leave for 45 mins + No final pressure drop = success

	Plastic (polybutylene) pipework Test B: + Fill system and pressurise to working pressure + Leave for 30 mins and check pressure + Leave for a further 30 mins and check pressure + If the pressure loss is less than 0.6 bar with no visible leaks = success Any pressure drop or leak requires rectification and retesting.

Commissioning	**Once the system has been tested for soundness and passed:**
	+ flush through the system with cold water – remove debris and flux
	+ refill system
	+ inspect water levels in cisterns
	+ inspect and check valves are holding (FOV, taps and drain offs)
	+ check flow rates at outlets (flow or weir cup) – manufacturers' instructions
	+ check performance at outlets:
	  + flow rate – (flow/weir cup)
	  + pressure test – (pressure gauge)
	  + temperature – (thermometer)
	+ manufacturers' instructions are followed and Benchmark certificate is completed
	+ verbal instructions from the customer and work colleagues need to be followed

Decommissioning	**Temporary decommissioning:** + Normally carried out while servicing or replacing components + Advise customer + Reinstated after a short period + Isolate mains water supply + Isolate electrical supply (safe isolation procedure described in Chapter 7) + Drain system or part of system safely **Permanent decommission:** + Normally carried out if the total system is removed + Not due to be reinstated + Isolate mains water supply + Isolate electrical supply (safe isolation procedure described in Chapter 7) + Drain system safely + Cap off incoming mains and label + Remove electrical supply from fused spur safely + Remove all component parts from system

Check your understanding

12 How much drop are you allowed on a soundness test?

Exam tip

Make sure you revise safe electrical isolation (described in Chapter 7 (Unit 211)). There might be hot water questions that come up in the exam about replacing electrical components, like showers and pumps.

Exam-style questions

1 Which of the following is a correct part of a soundness test?

a Pressurise the system to two times the working pressure

b Pressurise the system to three times the working pressure

c Pressurise the system to 1.5 times the working pressure

d Pressurise the system to atmospheric pressure

Exam tip

Remember from your workshop task what factors were used to calculate the test pressure.

2 What type of shower has been installed in this customer's property?

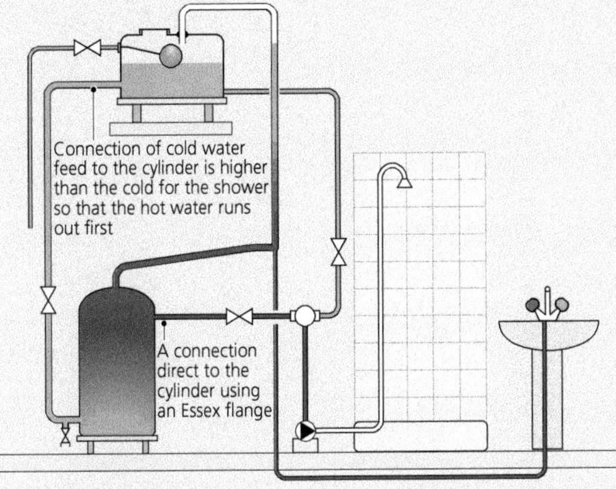

a Double impeller

b Single impeller

c Electric instantaneous

d Gravity

Exam tip

Look at the diagram in Question 2 carefully and follow the pipework from the CWSC to the shower outlet and note if there are any components on the way. Identify those items.

3 A customer complains that over a period of time, the spray from their shower head has reduced. What is the most likely cause?

a The head height of the cold water storage cistern has been reduced

b The hot water cylinder thermostat is faulty

c The pressure in the system has altered

d Limescale is building in the shower head

4 Which part of the hot water cylinder is the hottest?

a The bottom

b The middle

c The top

d The temperature is even throughout

5 How would you prevent the water being turned back on accidentally when temporarily decommissioning a system?

a Isolate using the supplier's valve

b Remove all pipework and cap off

c Label the valve 'DO NOT USE'

d Inform people in the property

6 Which of these is the correct sequence?

a Visual inspection – Soundness test – Pressure test

b Flush system – Visual inspection – Soundness test

c Pressure test – Soundness test – Visual inspection

d Visual inspection – Soundness test – Flush system

7 What type of shower controls the water temperature to the shower head, even when there are changes in the flow and temperature of the supplied water?

a High pressure mixer valve

b Lower pressure mixer valve

c Non thermostatic mixer valve

d Thermostatic mixer valve

8 Which is the key Building Regulations Approved Document for hot water regarding temperature?

a A b H c L d G

9 What type of water heater is inlet controlled?

a Single point water heater

b Multi point water heater

c Thermal store

d Combination boiler

10 Which one of the following is NOT a factor when designing a hot water system?

a Occupancy

b Building insulation

c Building size

d Available services

11 Which of the following offers guidance on hot water systems?

a Building Regulations Approved Document L

b The Water Supply (Water Fittings) Regulations

c Building Regulations Approved Document G

d BS EN 806

12 What is the biggest weakness when installing a combination unit?

a Poor flow rate

b Slow heat recovery time

c Potential blockages

d Poor pressure

13 Why is a sacrificial anode installed at the base of a hot water cylinder?

a Prevent electrolytic corrosion

b Prevent limescale build up

c Increase flow rate

d Increase pressure

14 Which Approved Document would you refer to when choosing a new hot water cylinder?
a G b L c H d A

15 What is the correct temperature that a TMV should be set to for a domestic bath?
a 38–40°C
b 41–44°C
c 55–60°C
d 47–50°C

16 When commissioning an appliance, which document should you look at to check the flow rate and pressure the appliance should produce?
a Building Regulations Approved Document G
b BS EN 806
c Manufacturer's instructions
d Building Regulations Approved Document L

17 Where is the expansion of the heated water taken up in a standard open vented hot water system?
a Up the cold feed and into the cold water storage cistern
b In the pipework
c Expansion vessel that is connected to the cold feed
d Expansion joints

18 Which one of the following systems is a centralised instantaneous hot water system?
a Single point water heater
b Unvented hot water system
c Combination boiler
d Open vented hot water cylinder

19 When decommissioning a system in a customer's property, which one of the following would you not carry out?
a Drain down
b Pressure test
c Electrical safe isolation
d Remove appliance

20 Which of the flowing is correct for the installation of a cylinder thermostat?
a Installed 1/3 of the way up a cylinder and set at 50°C
b Installed at the base of a cylinder and set at 60°C
c Installed at the top of a cylinder and set at 50°C
d Installed 1/3 of the way up a cylinder and set at 60°C

5 Central heating (Unit 216)

You need to familiarise yourself with central heating system layouts, component parts and the legislation covering central heating systems. Ask yourself:

- What legislation covers these systems?
- Why are central heating systems installed and what are the benefits to customers?
- What component parts make up a central heating system?
- What does each component part do within the system?

LO1 Understand central heating systems and their layout

Topic 1.1 Sources of information

REVISED

Statutory regulations:

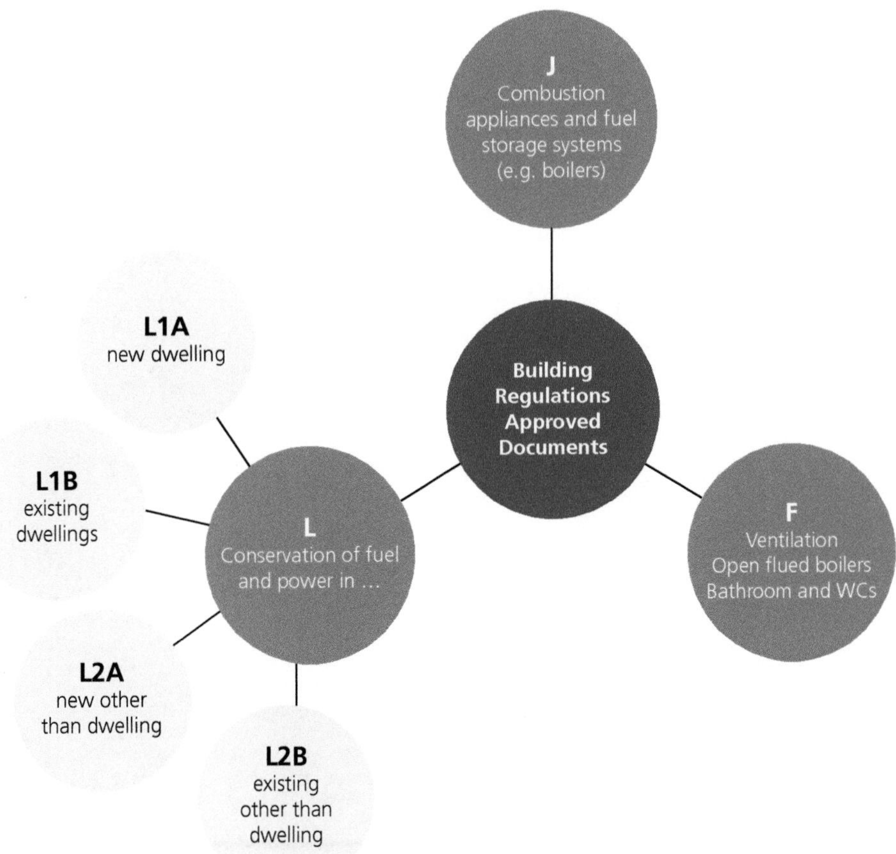

Industrial Standards:

- BS EN 12828
  - Heating systems in buildings.
  - Design for water-based heating systems.
- BS EN 14336
  - Heating systems in buildings.
  - Installation and commissioning of water-based heating systems.

- BS EN 442
  - Radiators and convectors.
- Domestic Heating Compliance Guide
  - Practical assistance for design and installation of systems to Building Regulations.

Manufacturers' instructions

- These must be followed to install, commission and maintain systems.
- They are always available online.

Check your understanding

1 Describe a double panel single convector radiator.

2 Which British Standard covers a double panel single convector radiator?

Now test yourself TESTED

1 You are having a new extension added to your house. Which part of the Building Regulations would the central heating system come under? Explain why it is important to follow the Building Regulations (Approved documents).

Typical mistakes

Many students struggle to recall the correct regulations and standards, so it is worth trying to memorise the Building Regulations chart for the exam.

Topic 1.2 Operating principles of systems and components

REVISED

This section will cover various terms and systems that can be found and are installed in customers' properties.

It is particularly important for you to understand this chart, to get an overview of central heating systems, especially under the titles 'open vented' and 'sealed'.

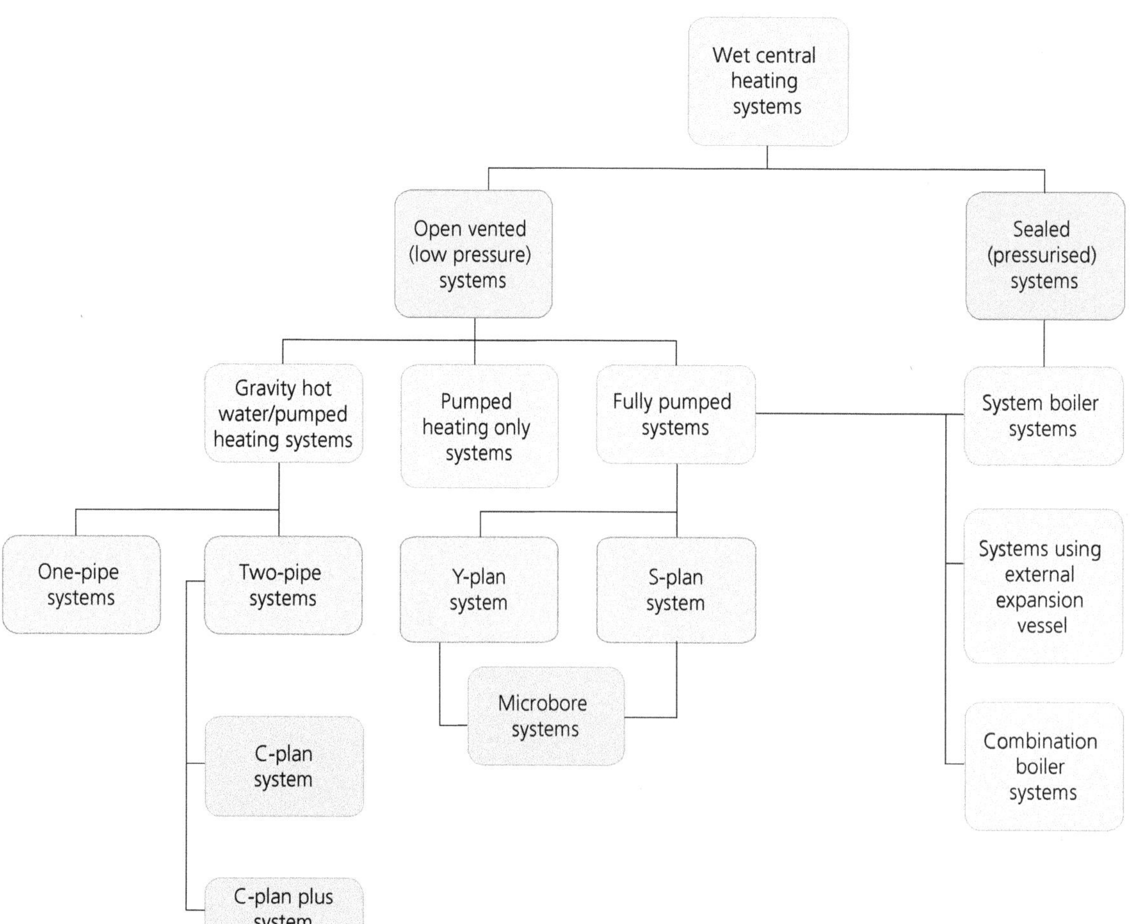

Central heating systems

Types of systems

The three definitions below broadly cover the options for heating within a property.

Water central heating:

- Primary water is used to convey heat around the property to radiators or convectors.
- Boiler heats up the primary water.
- Fuels for the boiler can be electricity, gas, LPG, oil or solid.

Warm air:

- Air is heated to convey heat around the property to vents located in the rooms.
- Warm air heater heats up the air which is forced around duct work.
- Fuel is generally gas.

Storage heaters:

- These are localised in each room.
- Thermal bricks are heated electrically overnight when cheaper rates are offered.
- Storage heaters release the heat when required during the day.

Heat emitters

Panel radiators:

- Most commonly installed.
- Today, panel radiators must have convectors.
- Made of pressed steel.
- Most common radiator height is 600 mm.
- Rely on convection currents to heat the room up.
- Convectors add more surface area to help heat the room more quickly.

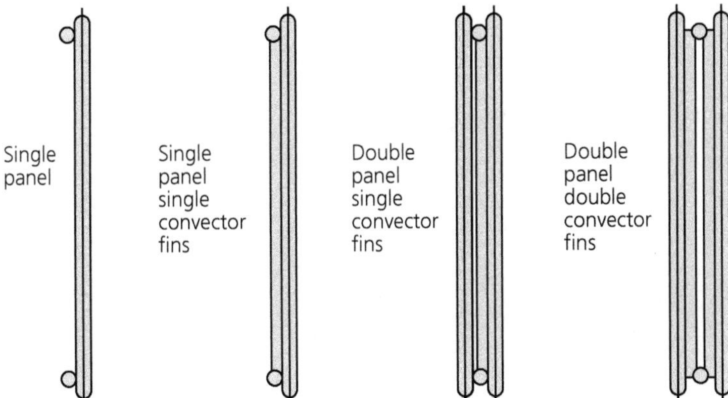

Panel radiators

Column radiators:

- Also known as 'hospital' radiators.
- Each column used to be connected individually – more columns means more heat.
- Used for period refurbishment.
- Made from cast iron or aluminium.
- Rely on convection currents to heat the room up.

Low surface temperature radiators:

- A convector or radiator with a surface cover.
- Surface temperature does not go above 43°C.
- If anyone touches the radiator, they do not get burnt.
- Used in hospitals, care homes and nurseries.
- Rely on convection currents to heat the room up, with colder air entering at the bottom, releasing warmer air from the top.

Low surface temperature radiator

Fan convectors:

- Contain a finned copper heat exchanger and an electric fan.
- The fan draws in cold air and forces it through the heat exchanger, blowing warm air out the top.
- This warms up larger areas more quickly.

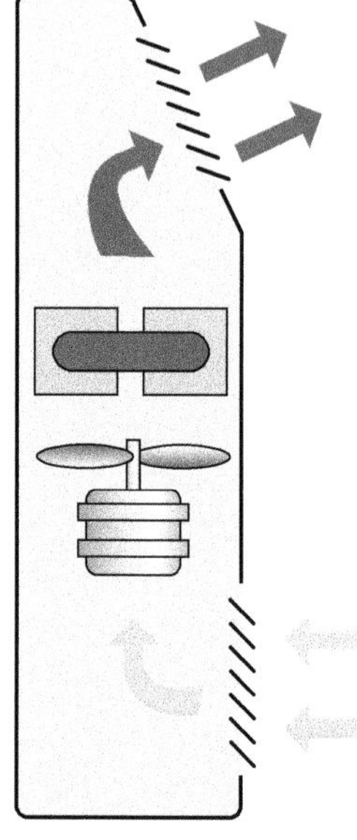

Fan convector

Plinth heaters:

- Also known as 'kick space' heaters.
- Designed to be installed below kitchen units.
- Also found in vanity units and bottom of staircases.

Towel warmers:

- Installed in bathrooms.
- Designed in a variety of styles.
- They can incorporate an integral radiator.

Connections:

- **TBOE** – top bottom opposite ends.
- **BBOE** – bottom bottom opposite ends.
- **TBSE** – top bottom same end.
- **BBOE** is commonly used in properties today.
- The bottom of the radiator needs to be at least 150 mm off the floor to allow sufficient circulation of air (convection).

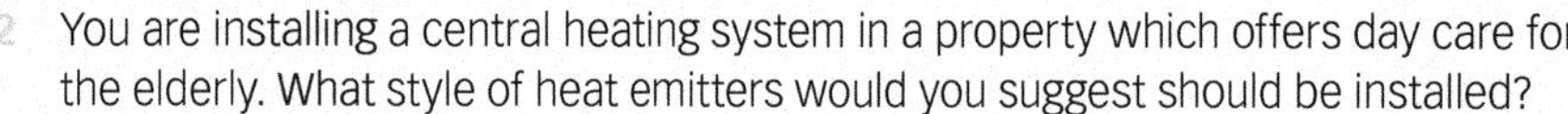

Now test yourself

TESTED

2. You are installing a central heating system in a property which offers day care for the elderly. What style of heat emitters would you suggest should be installed?
3. The local community hall is having a refit. They have asked your opinion on what style of heat emitter would heat the hall up quickly. What would you suggest?
4. A primary school requires an additional heat emitter. What style should be installed?

Heating components

Radiator valves:

1. **Wheel head:** These allow manual control for the radiator to be turned on/off. In modern systems, the wheel head valve is replaced by a thermostatic radiator valve (TRV).
2. **Lockshield:** This is basically a wheel head valve with a secured cap. Set by the plumber to regulate the flow and return so the system is 'balanced'.
3. **Thermostatic radiator valve (TRV):** Automatically controls the room temperature by regulating the flow and return. They react to air temperature. These must be fitted on all new installations (part L) and every radiator (except the radiator where the room thermostat is located).

Automatic air vents:

- Installed at the high points in the system.
- They allow any build-up of air to be automatically emitted from the system.

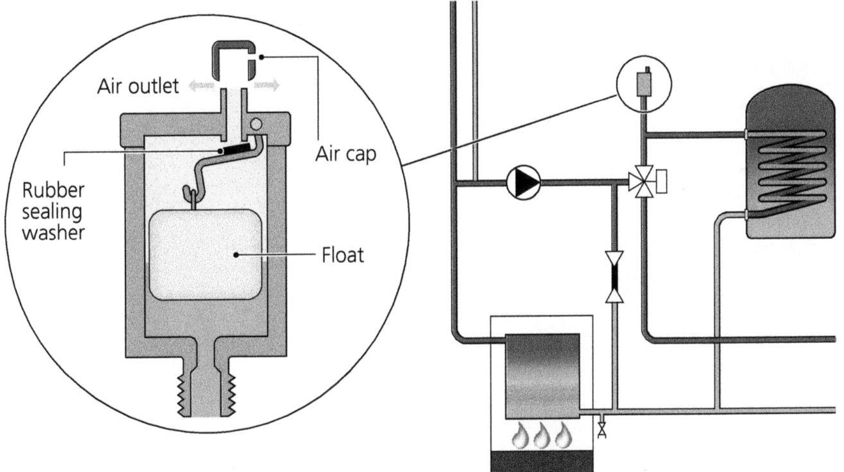

Automatic air vent

Expansion vessel:

- Fitted to a sealed system and replaces the feed and expansion cistern.
- It allows the 4% expansion of hot water safely.
- It is a steel cylinder divided by a rubber diaphragm.
- Dry side is pressurised.
- Wet side expands as the water is heated.
  - A – at rest with water cold.
  - B – water heating up.
  - C – water at temperature.

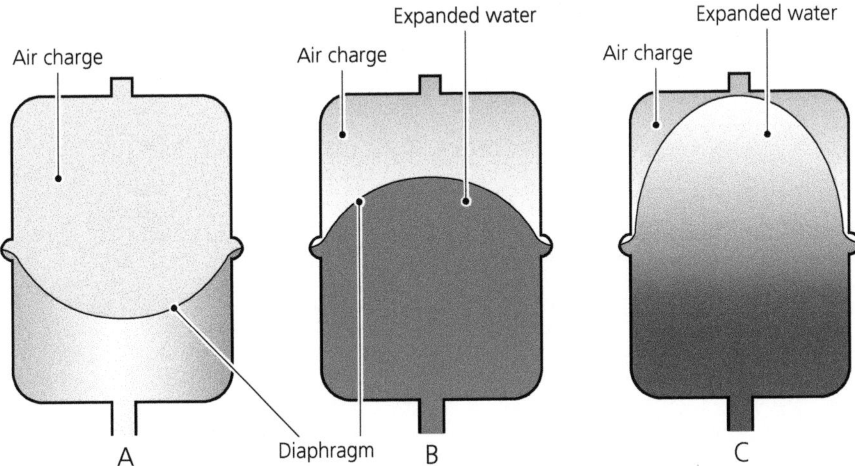

Expansion vessel

Filling loop:

- Fitted to sealed system and replaces the cold feed to the primaries.
- It connects the mains cold water (Category 1) to the central heating water (Category 3).
- Mains protected by double check valve.
- 'Temporary filling loop' must be removed after filling.

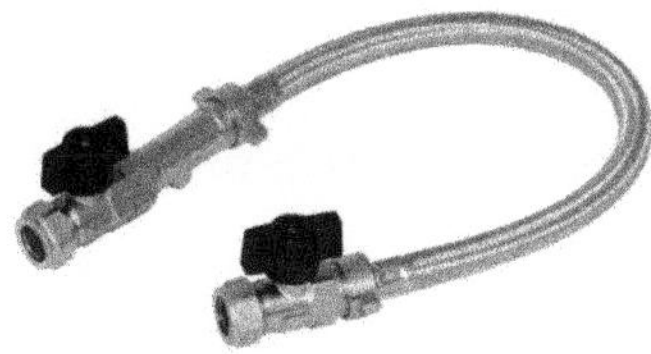

Filling loop

Pressure gauge:

- Fitted to a sealed system.
- It identifies the filled pressure and the working pressure of the system.

Motorised valves:

1 Two port – S plan system
   - One valve controls the flow of water to the hot water cylinder (controlled by the cylinder thermostat).
   - One valve controls the flow of water to the central heating system (controlled by the room thermostat).
   - They isolate a circuit when closed, so also act as a zone valve.
   - The paddle is driven by a syncron motor.

Syncron motor An electric motor located inside a two-port valve that moves the paddle and engages the micro-switch. These can be replaced easily.

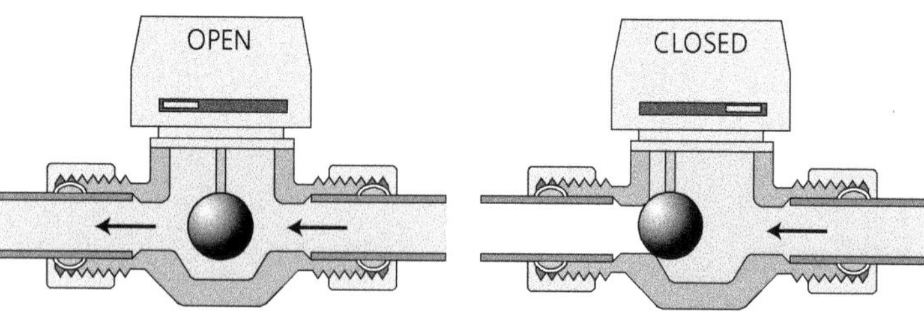

Two port valves

2 Three port mid-position – Y plan system
   - This one valve controls the flow of water to both the hot water cylinder and the central heating system, via a paddle.
   - Resting position is hot water open.
   - Can open either or both circuits.
     - Port A – Central heating.
     - Port B – Hot water.
     - Port AB – Flow from the boiler.

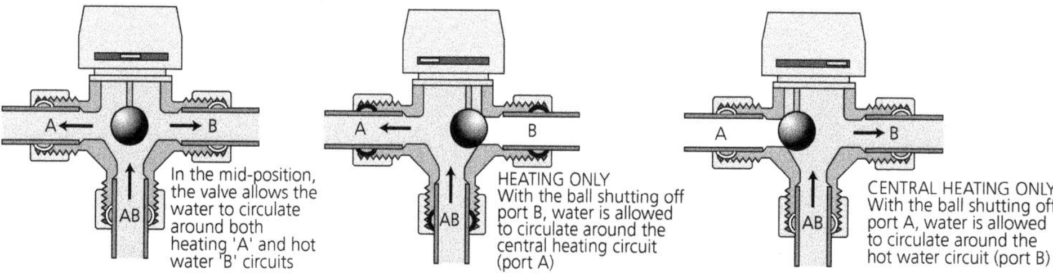

Three port mid position valves

3 Three port diverter – W system or a combination boiler
   - This is hot water priority. Until the hot water gets to temperature, the central heating circuit will not open.
   - Only one circuit can open at any one time

Feed and expansion cistern:

- Installed on an open vented system.
- Fills the system with water and maintains water level.
- Allows expansion of heated water.
- Open vent from the primaries terminates over the F&E.

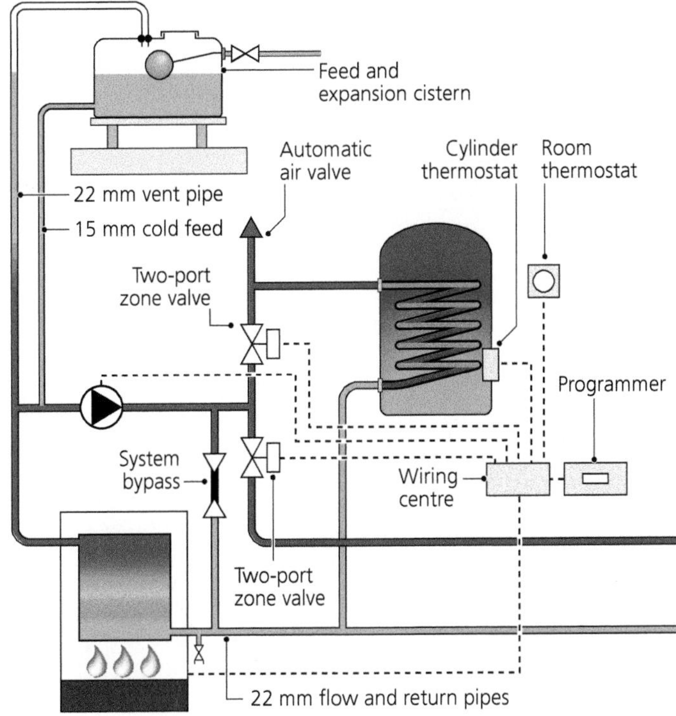

Open vented system with feed and expansion cistern

Circulating pump:

- Positioned with care to avoid corrosion and aeration.
- Electric motor with an impeller.
- Circulates the heated water around the system and through the boiler.
- A central heating system must have 'positive' pressure.

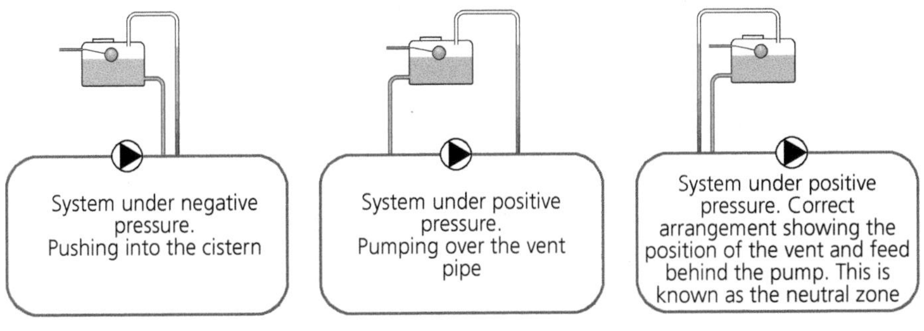

Pump position

Impeller An internal rotating paddle that powers the water in a pump.

Exam tip

If you are asked a question about circulating pumps, remember they must always draw on the cold feed. In other words, a circulating pump is fitted after the cold feed to give positive pressure to the system.

(NB: The pump position is also listed in LO1.4 but will not be referred to again.)

Automatic bypass valves:

- Control the flow of water between the flow and return if other routes are closed.
- The pressure builds up and the automatic bypass starts to open.
- Reduces possible noise and boiler lock out.

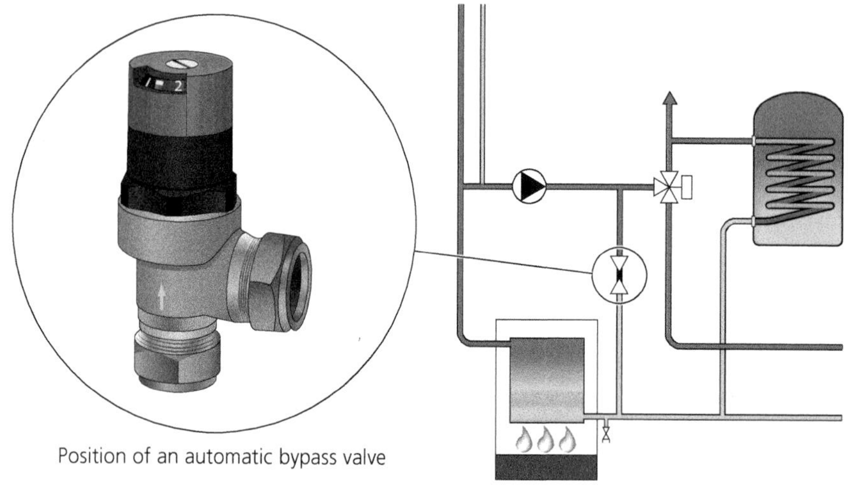
Position of an automatic bypass valve

Automatic bypass valve

Thermo-mechanical cylinder control valves:

- Non-electrical control fitted on old systems to control water temperature in the hot water cylinder.
- This is the minimum standard to comply with the Domestic Heating Compliance Guide and Building Regulations Part L1B.

Anti-gravity valve:

- Non-electrical control fitted on old systems to prevent any unwanted gravity circulation.
- It is similar to a single check valve.

Drain off valve:

- Installed at every low point in the system.
- Allows part or total drain down of the system.
- It can be soldered, press fit, push fit or threaded.

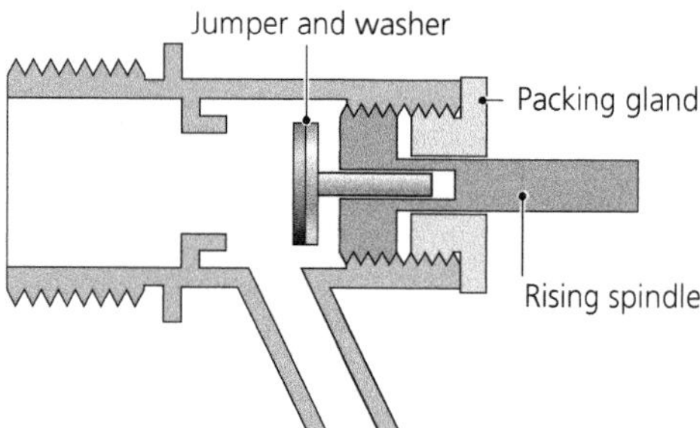

Drain off valve

Additives:

- Corrosion inhibitor must be added to comply with warranty.
- Stops black sludge – magnatite.
- Reduces fuel cost.
- Prevents hydrogen gas build ups, pinholing in radiators and limescale.

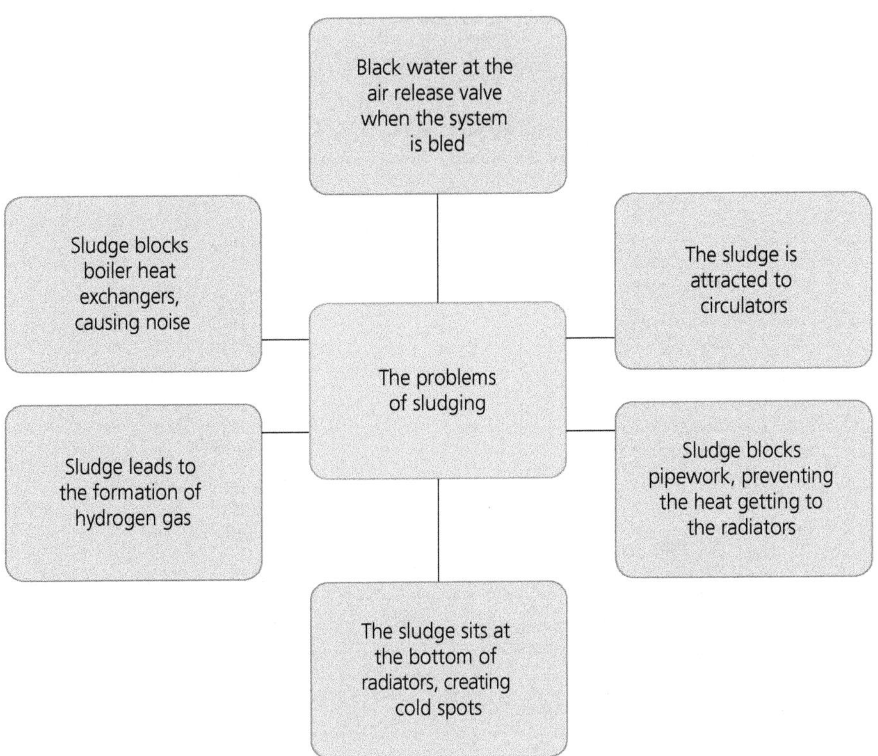

Sludge problems

Low loss header:

- Allows the system circulation to remain constant for the boilers by dividing a larger system into a primary and secondary circuit.
- **Primary circuit** – constant for the boilers.
- **Secondary circuit** – allows fluctuations in demand to different circuits.

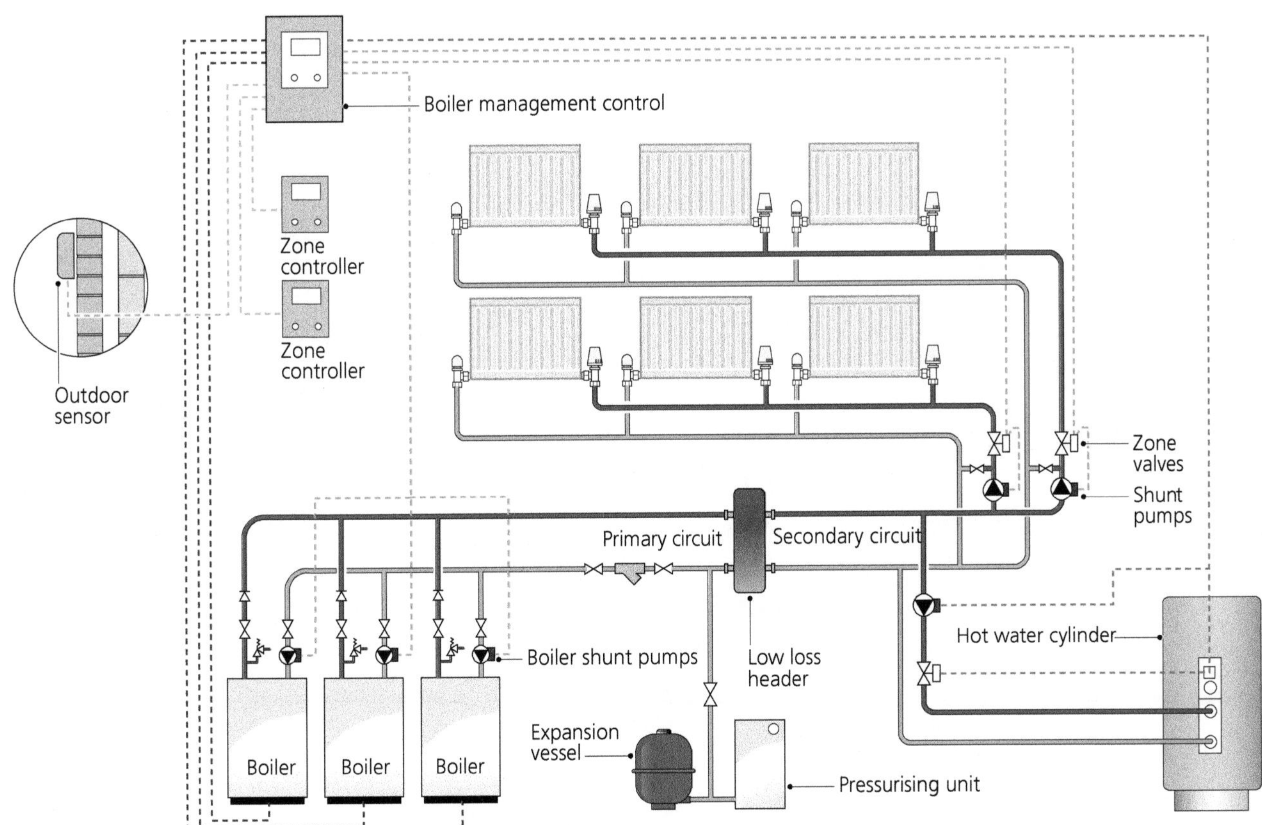

Low loss header

Buffers:

- Buffer tanks hold hot water ready to be circulated around a system when required.
- Store hot water when demand is low and release it when demand is high.
- Often connected to renewable sources, like heat pumps.

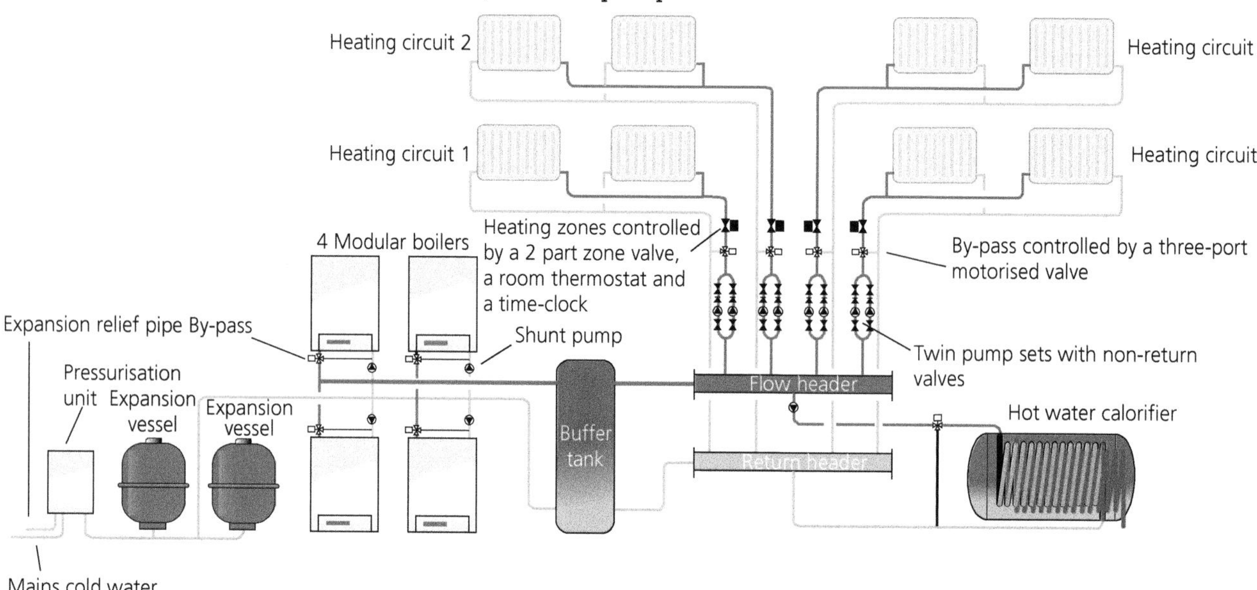

Heating system containing a buffer tank

Pressure relief valves:

- Installed on a sealed system and combination boiler.
- Safety valve if the expansion vessel fails.
- Drip to outside.
- Pre-set to about 3 bar.

Expansion joints:

- Used on larger installations where there are long straight sections of pipe.
- Protect from distortion.
- Can be pipework loops or bellows type.

Corrosion filters:

- These are magnetic or centrifugal.
- They collect the black sludge – magnetite.
- Prevent blockages, corrosion build up and hydrogen gas build up.
- Often need installing as part of a new boiler warranty.
- Need emptying every boiler service.

Exam tip

If you are asked in an exam question about the neutral point on a central heating system, remember it is at the base of the cold feed.

Check your understanding

3. Which central heating valve allows the installer to 'balance' the system?
4. What valve must be installed at every low point in a central heating system?
5. Why is it important to install the central heating circulator in the correct position?
6. If an expansion vessel on a sealed system failed, what would be the first symptom?
7. Between which two pipes in a central heating system is an automatic bypass valve installed?

Typical mistakes

If you are asked about component parts of a central heating system in the exam, try to remember the circuit diagrams and where each component is located. That should give you a strong indication of the function.

Now test yourself

TESTED

5 A customer calls you to their property because their system is losing pressure regularly. They have noticed dripping from a pipe on the outside wall near to the boiler when the system is hot. What valve is allowing the dripping?

Types of boilers

Table 5.1 Types of boilers

Condensing: + All boilers fitted today must be condensing, high-efficiency boilers + A rated – Building Regs Part L + 93%+ efficient + Flue gas cools to form 'pluming' + Two heat exchangers to extract more heat for the system + Requires a condensate drain (slightly acidic)	**Combination:** + Space saving + Used in smaller properties + Supplies central heating and instantaneous hot water + Hot water priority
Freestanding: + These are floor mounted boilers + Tend to be older open flued boilers + Commonly replaced with wall mounted	**Wall mounted:** + These are common domestic boilers mounted on the wall + Often in a kitchen cupboard + Older styles can have cast iron heat exchangers + Modern styles have copper or aluminium heat exchangers + Burner under heat exchanger with flue to outside
Open flued: + Old type of boiler + Care required due to possible carbon monoxide build up + Draws air from the room the boiler is installed in, which in turn, must have an air vent to outside + Vertical flue is separate to extract combustion gases	**Room sealed (natural draught):** + Older boilers + Circulation of air in and flue gases out is carried out by natural air movement + Air in a flue gas extraction is from the same terminal on the outside wall

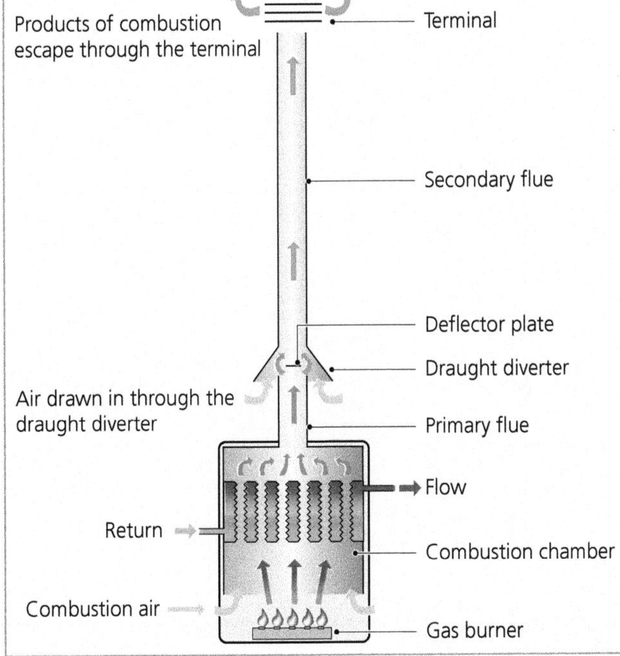

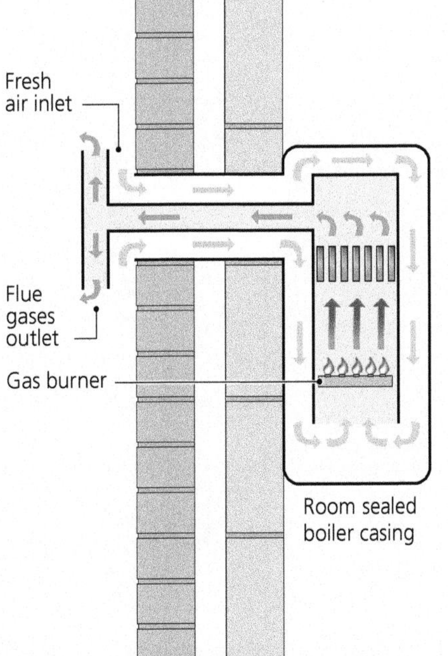

Room sealed (fan assisted):

- Like natural room sealed boilers, except a fan forces air in and combustion gases out.

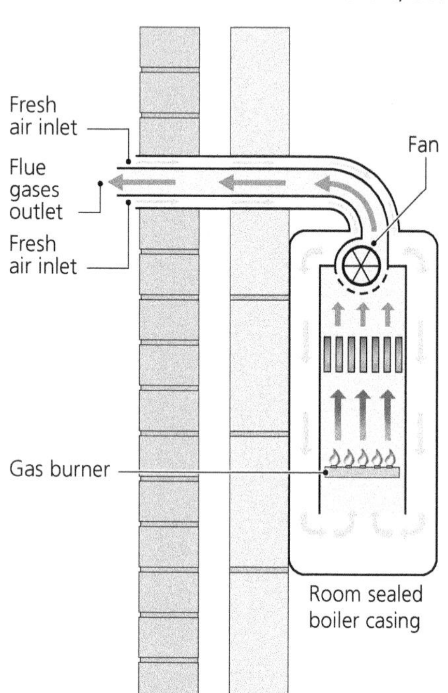

Room sealed Where the boiler draws air direct from outside of the building through the same flue used to discharge the combustion gases.

Check your understanding

8 Which statutory regulation would cover the installation of A-rated boilers?

Topic 1.3 Filling and venting systems

REVISED

In this section we will be looking at the system styles and pipework layout. It is important to be able to follow the pipework with your finger, identifying components on the way. This will help you with exam questions that ask about systems and their components

Types of systems

Pumped heating and gravity hot water:

- Also known as a 'C plan+' system.
- A circulator forces water around the central heating system (quick heat up).
- Convection (gravity) circulates the primary water to the cylinder (slow heat up).
- Minimum accepted system under Building Regulations Approved Document part L1B.
- Limited control:
  - one-off two port valve
  - room stat
  - cylinder stat
  - programmer
  - pump to CH.

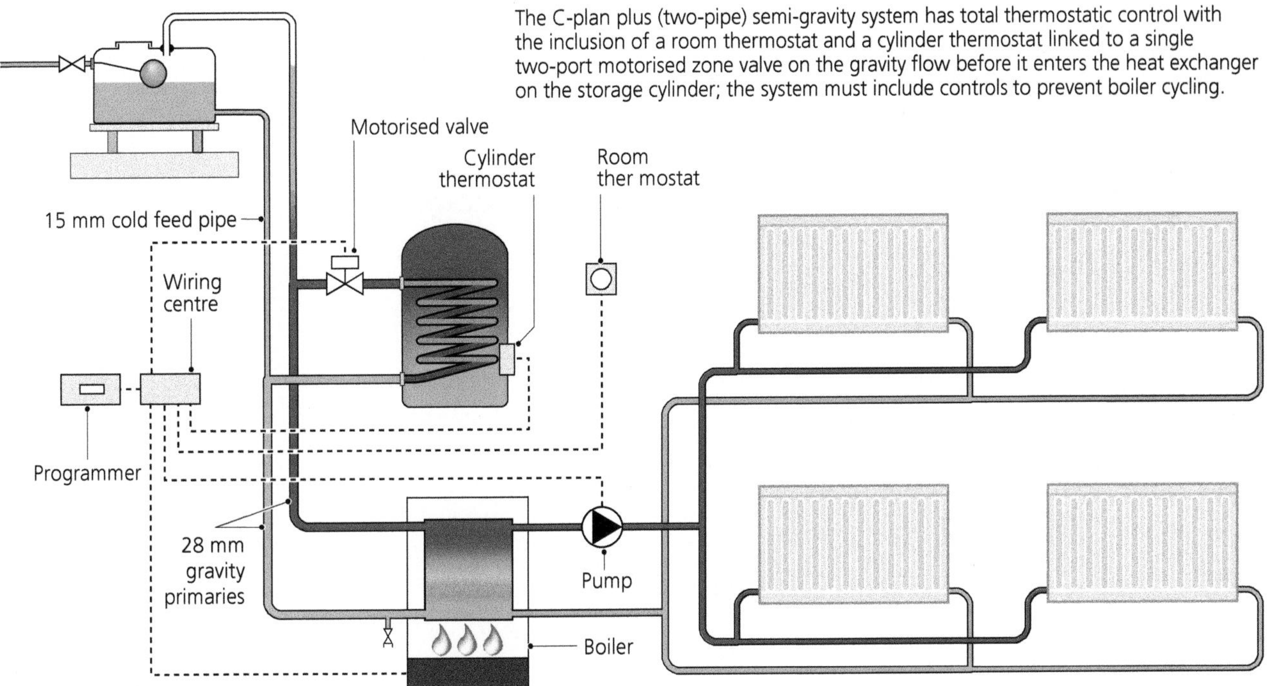

Pumped heat gravity water

Fully pumped (S plan):

- Has two × two port valves: one for central heating, one for hot water.
- Zone valve isolates the circuit when closed.
- Both sides pumped.
- Full control:
  - two port zone valve
  - cylinder stat
  - room stat
  - programmer
  - auto bypass
  - pump
  - TRVs.

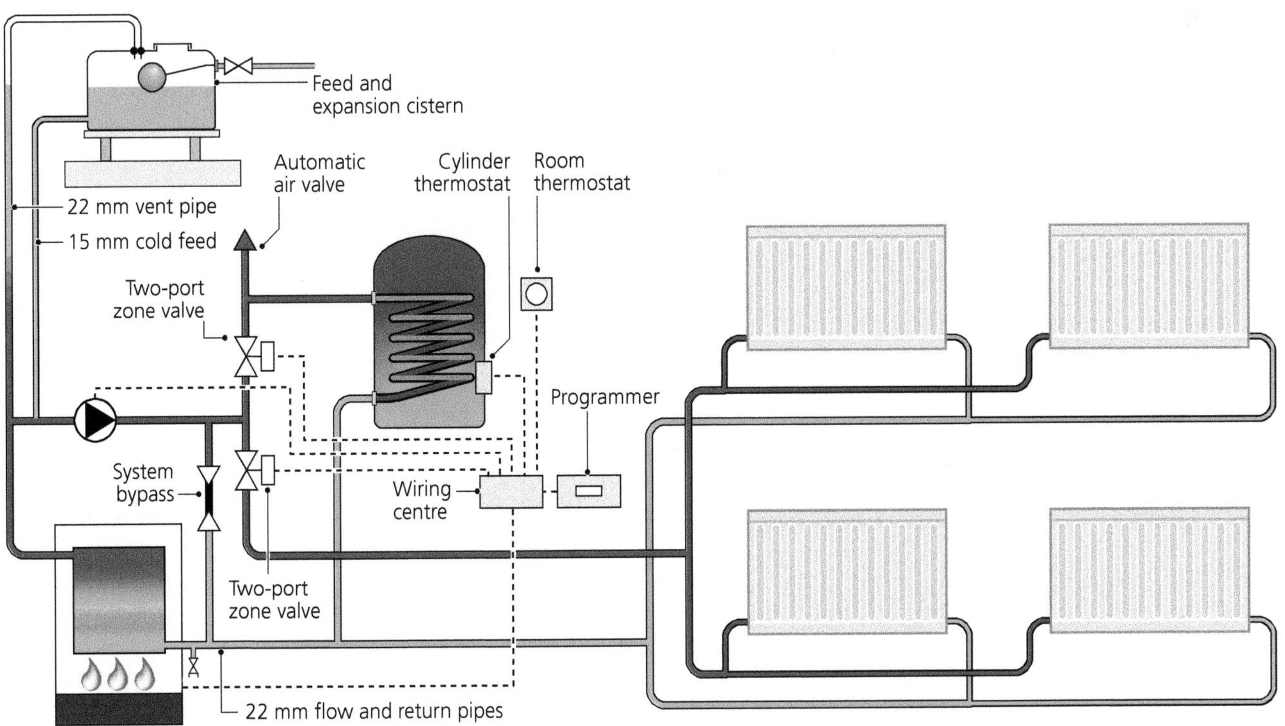

S-plan

Fully pumped (S plan +):

- In properties bigger than 150 m^2.
- This adds additional zone valves to the system.
- Zoning upstairs and downstairs – heating separately.

Fully pumped (Y plan):

- One-off three port valve, controlling the flow of water to one or both circuits.
- Both sides pumped.
- Full control:
  - one × three port valve
  - cylinder stat
  - room stat
  - programmer
  - auto bypass
  - pump
  - TRVs.

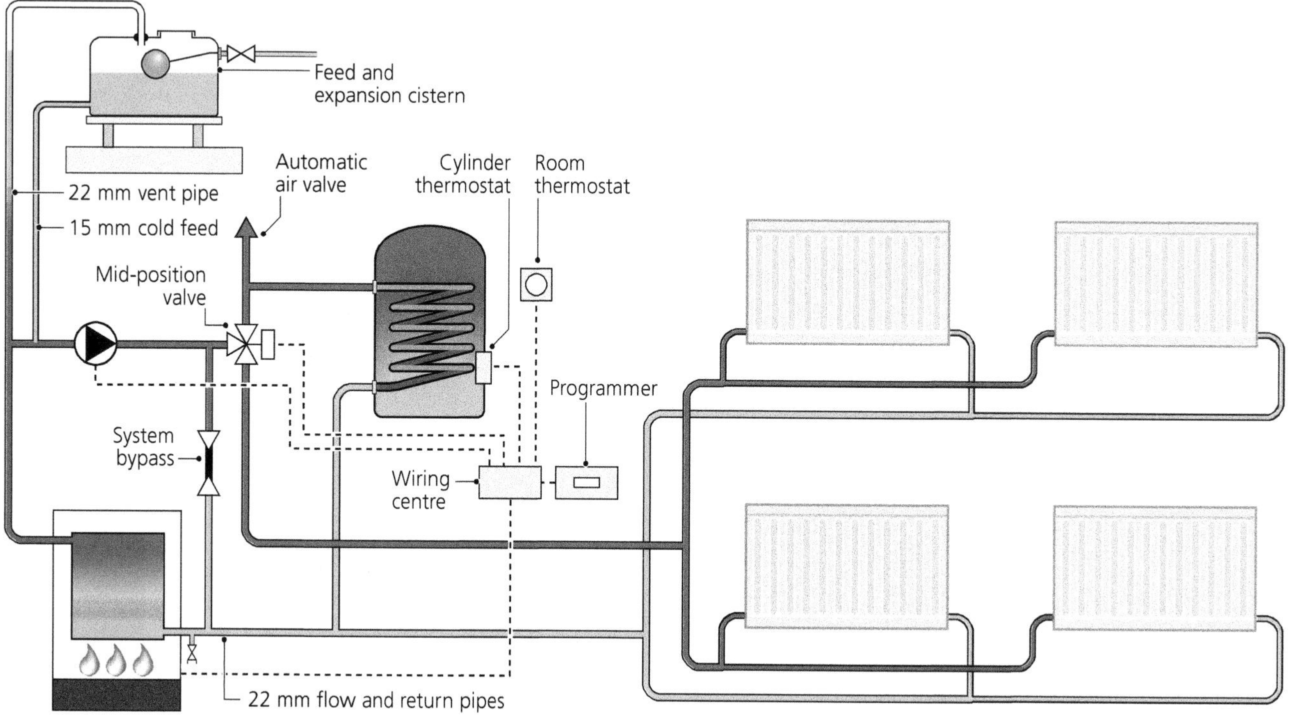

Y plan

One pipe system:

- Old system – not used.
- System is not to Building Regulations Approved Document Part L standards.
- Lack of temperature control.
- The radiator water cools as it travels from one radiator to the next.
- Last radiator in the circuit takes a long time to heat up.

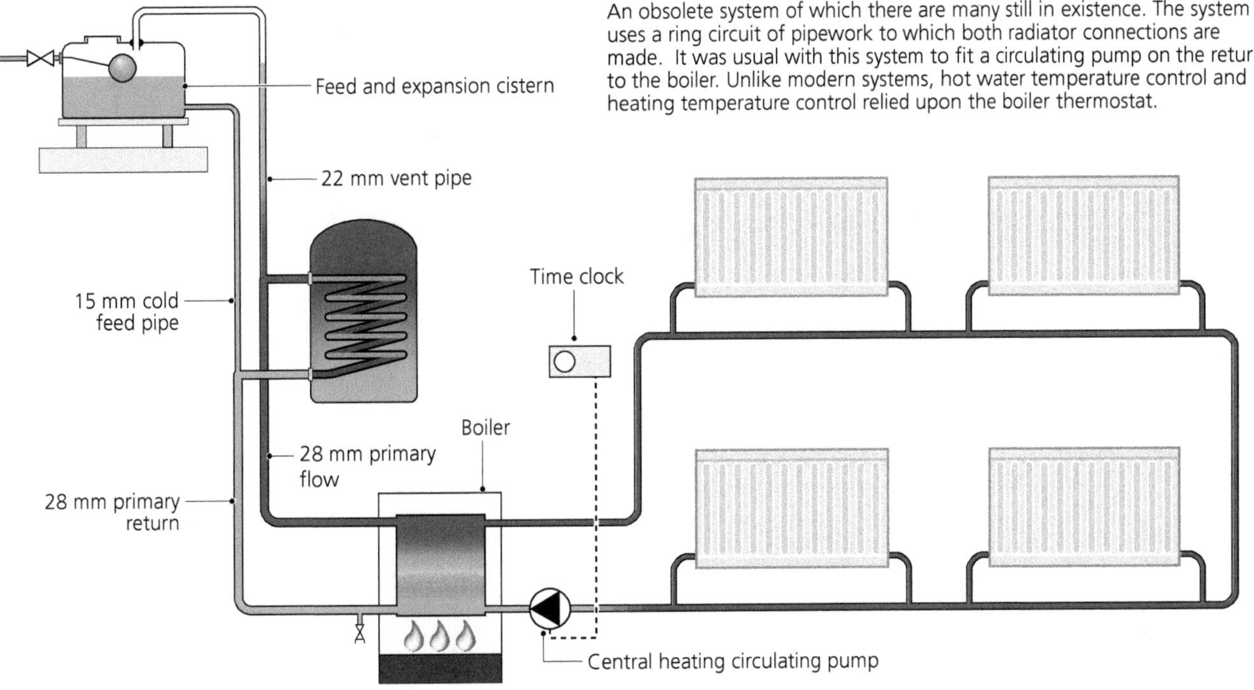

One pipe

Two pipe system:

- This system shown is the early two pipe system.
- Separates the flow and return pipework.
- Used in the S and Y plan systems today, which are to Building Regulations Approved Document Part L standards.

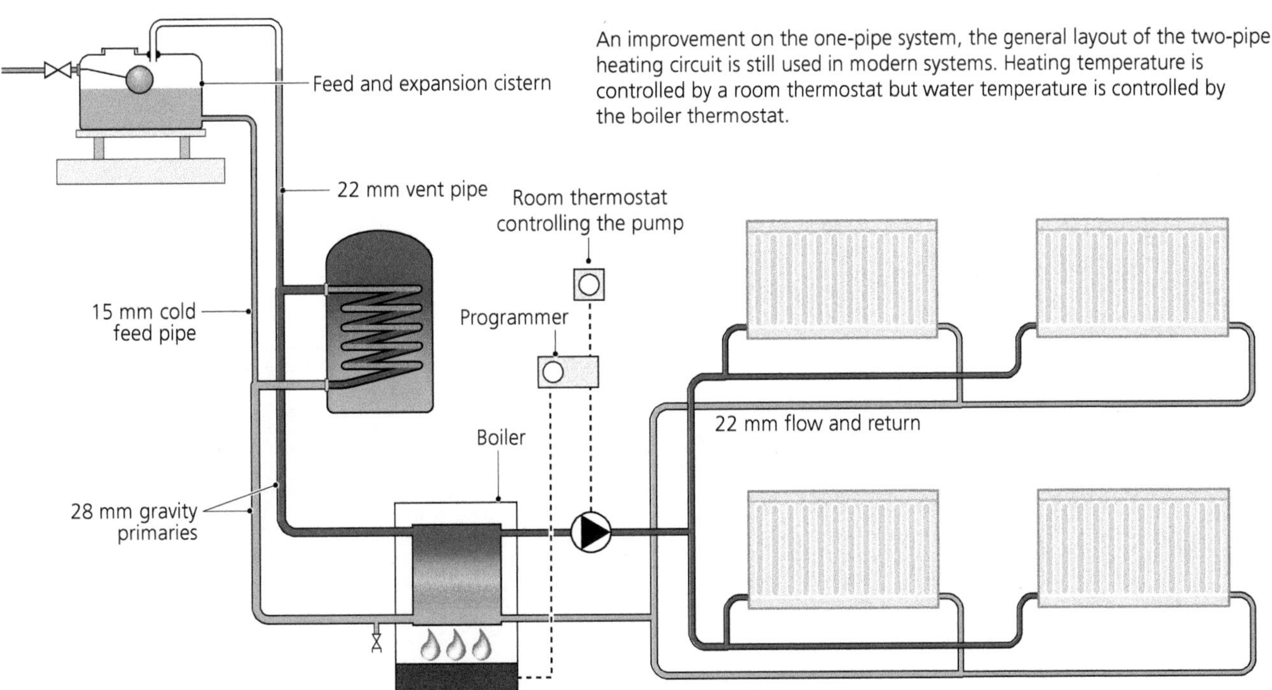

Two pipe

Manifold:

- Use in microbore (or minibore) systems.
- A multi-connection fitting or manifold, installed to the flow and return pipe.
- Heat emitters are fed by 8 mm or 10 mm pipe instead of 15 mm.
- Prone to blocking (black sludge).
- More complex manifolds are used in underfloor heating systems.

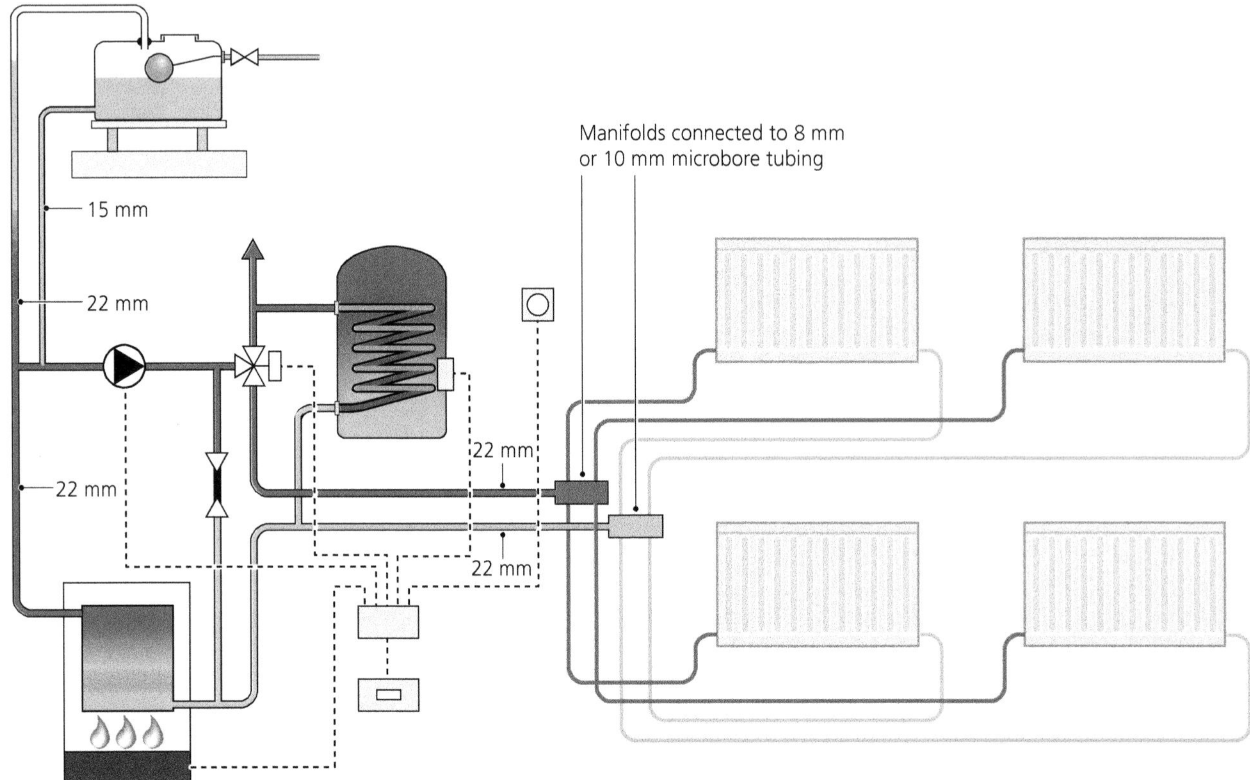

Microbore system

Underfloor heating:

- Modern effective heating method.
- Floor becomes the heat emitter.
- Low temperature method of heating.
- Can be attached to a renewable source.
- Flow temperature of 40–45°C.
- (Central heating flow temperature 70–80°C.)
- Individual circuits for each room leading back to a manifold.
- Each room has a room thermostat.

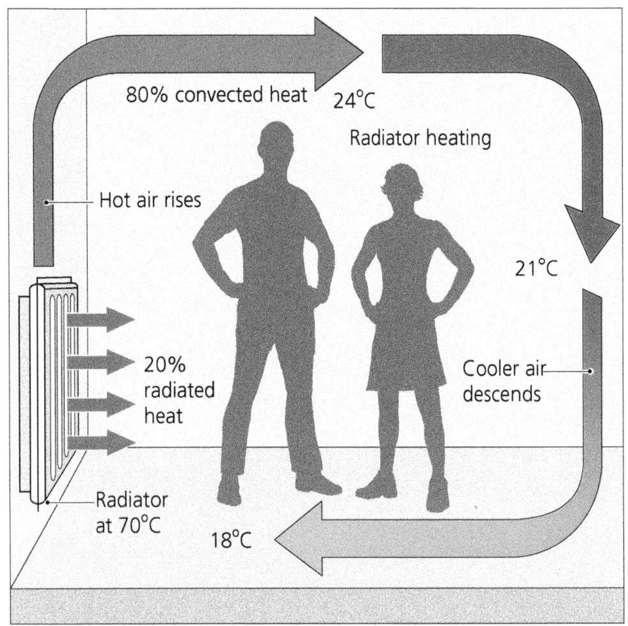

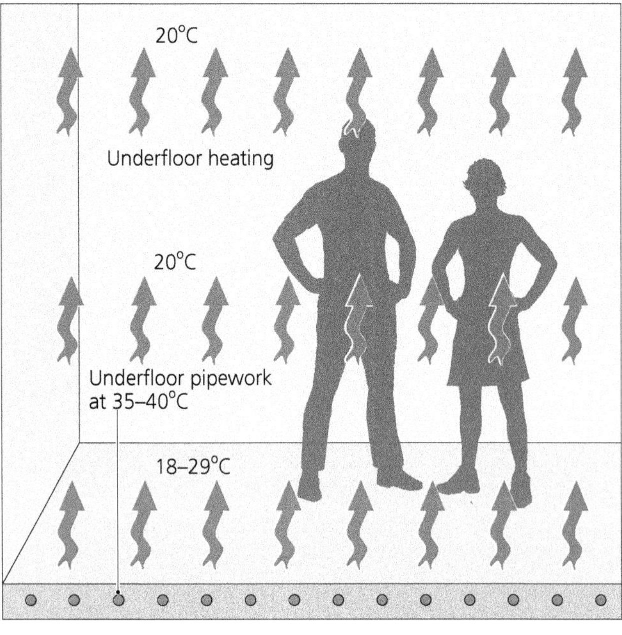

Radiator heating compared with underfloor heating

> **Exam tip**
>
> In an exam you might be asked to explain the function of the primary flow and primary return pipes or to state what they are connected to. The **flow** pipe takes the heated water from the boiler to the radiators, the **return** pipe takes the cooled water back to the boiler for reheating.

> **Exam tip**
>
> Follow the system pipework with your finger to trace the flow of water from the boiler and identify each component on route.

Topic 1.4 Filling and venting of systems

Once the system has been completed, it will need to be filled with water and the air must be released. Look at each style of system and revise how to outline the specific areas of a system, and then how to fill and vent each system. Refer back to LO1.2 for the pump position.

There are features to allow for appropriate filling and venting of systems: open and sealed.

Filling and venting: open vented system

Feed and expansion cistern:

- It is located at the top of the system. Open vent from primaries must terminate over the feed and expansion cistern.
  - Water enters the system.
  - Allows for expansion.
  - Provides a static head of water.
- It is 18 litres in size.
- Open vent = 22 mm (min).
- Cold feed = 15 mm (min).

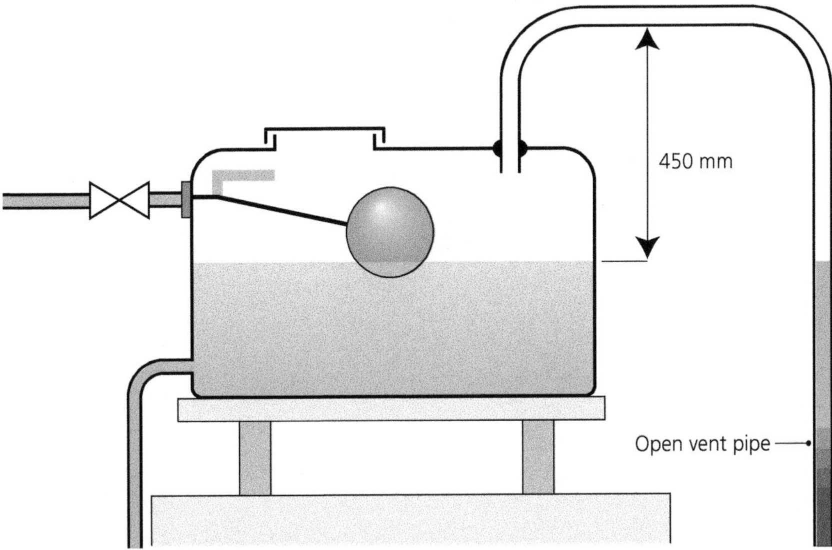

Feed and expansion cistern

Cold feed and open vent connections:

- Two options of design to release air.
- H frame connection
  - Open vent (22 mm) a maximum of 150 mm apart from the cold feed (15 mm).
  - Pump draws on the cold feed.
  - Neutral point of the system is at the base of the cold feed.
- Air separator
  - Open vent (22 mm) and cold feed (15 mm) ready-made connections to separator.
  - Pump draws on the cold feed.
  - Neutral point of the system is at the base of the cold feed.
- Both allow any air in the system to rise up the open vent.

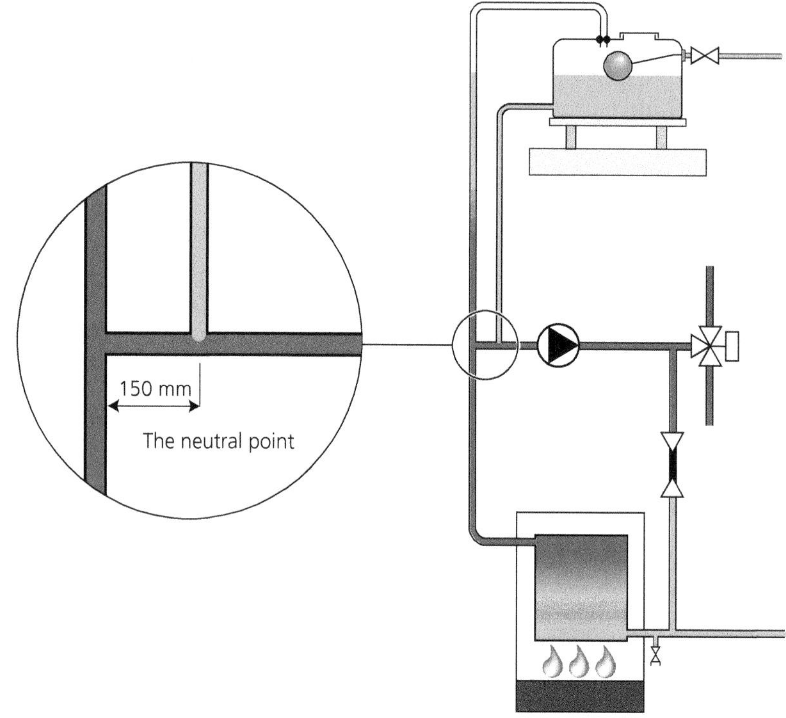

H frame

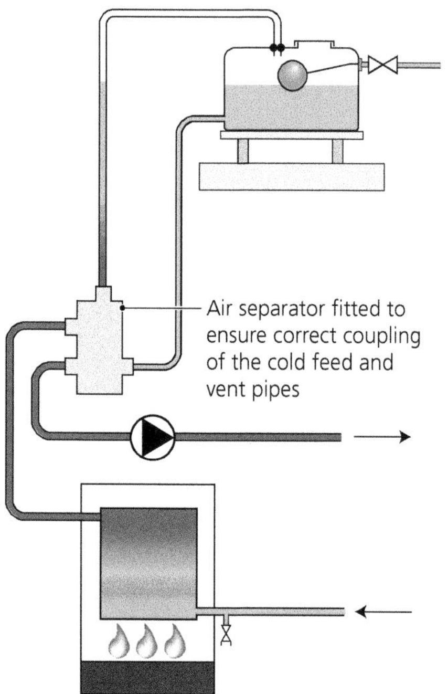

Air separator

Releasing air from the system:

- Air is released from any high point in the system.
- When filling the system, the radiators will need to be bled while the majority of air will be released via the open vent. Any air vents will also need to be checked.
- When operating the automatic air vents, they will release air along with the open vent. Occasionally, the radiators may need bleeding.

Now test yourself TESTED

6 A central heating system has 150 litres of water in it. How much water does the F&E need to accommodate when the system is heated?

Filling and venting: sealed system

Expansion vessel, pressure gauge, pressure relief valve and filling loop:

- All installed on the return pipework near to the boiler.
- Temporary filling loop needs to be removed after filling is complete.
- Mains protected by a double check valve.
- Pressure gauge showing the system pressure.
- Pressure relief valve protecting the system against a rise in system pressure or failure of the expansion vessel.
- Expansion vessel taking up the expansion of heated water.

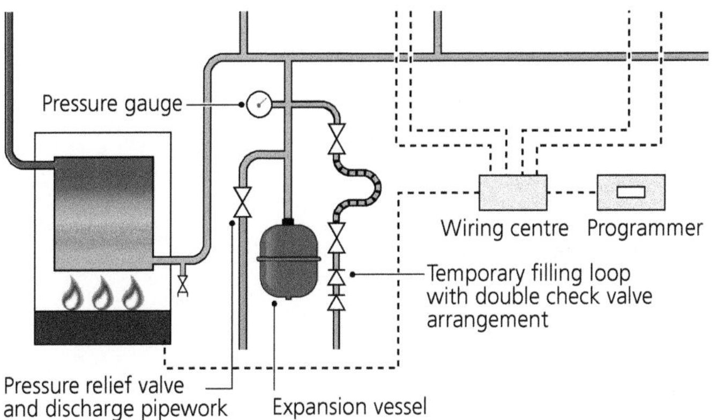

Filling and venting a sealed system

Releasing air from the system:

- As mains water enters the system, the pressure rises.
- As each radiator is bled, the system pressure reduces.
- Automatic air vents release air when operational and occasionally radiators may need to be bled.

Check your understanding

9 On a sealed system, what back flow protection device must be installed to protect the mains water?

Now test yourself TESTED

7 You are called to a customer's property where they have seen drips outside the house coming from a pipe near to the boiler. The boiler is not working now as the system pressure is too low. What do you suspect may have happened?

Topic 1.5 Types of fuels

There are different types of fuels used for heat producing appliances.

Solids:

- Hetas registered.
  - Coal
  - Wood
  - Pellets
- Tend to be used in rural settings.
- Large store area required.
- Access for lorry deliveries.

Gases:

- Gas Safe registered.
- Natural gas is:
  - most common fuel
  - supplied via the National Grid.

- LPG is:
  - supplied in bottles or large tanks
  - used in caravans, marine, hospitality and rural areas.

Oil:

- OFTEC registered.
- Tank required.
- Smelly.
- Rural areas.
- Access for lorry deliveries.

LO2 Install central heating systems and components

This is a workshop activity in which you will need to show that you know the methods used to install central heating systems and components, including the following:

- positioning and fixing requirements of pipework
- effects of expansion and contraction
- connecting to existing systems
- soundness test requirements for pipework
- installing and testing components
- replacing defect components.

Visual inspection, soundness test and commissioning procedures are the same as outlined in Chapter 3, Cold water (Unit 214), but for commissioning, the temperature of the water must also be tested.

LO3 Understand the decommissioning requirements of central heating systems and their components

Topic 3.1 Decommissioning systems

REVISED

The types of decommissioning methods are:

- **permanent decommissioning**: when a system is being taken out of service, dismantled or stripped out
- **temporary decommissioning**: when a system is being worked on for a short period of time (for example, replacing a part).

Topic 3.2 Preparing for decommissioning

REVISED

The methods to prepare for decommissioning systems are:

- keep customer and colleagues informed
- safely isolate electrics locally or at the consumer unit
- safely isolate water locally or at the inlet
- put up warning notices
- supply alternative heating if required (for example, for the elderly and infirm).

Topic 3.3 Decommissioning central heating systems

REVISED

Procedures to follow when decommissioning systems include:

- inform people
- isolate the services – gas, water and electricity. Isolating them locally is preferable for temporary isolation
- put up warning notices at isolation points
- use temporary continuity bonding when removing components or sections of pipe
- drain system safely in foul water sewer due to chemicals and sludge
- cap any open ends in pipework.

Exam-style questions

1 What type of central heating system is filled under mains pressure?
- a Sealed system
- b Open pipe system
- c Gravity system
- d Open vented system

2 Which of the following is the Building Regulations Approved Document for energy efficiency of the central heating system?
- a H
- b P
- c A
- d L

3 In which of the following locations is insulation of central heating pipework important?
- a Bedroom
- b Internal wall
- c Loft area
- d Kitchen

4 Where is the best place for an automatic bypass valve to be fitted?
- a The flow and return immediately above the boiler
- b The primary return after the circulator
- c Between the flow and return near the cylinder
- d Between the flow and return of the index radiator

5 What material is the plastic pipe used in central heating systems made from?
- a MDPE
- b Polybutylene
- c ABS
- d Polystyrene

6 Where does the water come from on the 'AB' connection of a three-port valve?
- a From the boiler
- b From the radiators
- c From the return pipe
- d From the cylinder

7 Which type of central heating system has the minimum controls to be compliant with Building Regulations Approve Document L1B?
- a S plan
- b Y plan
- c C plan +
- d W plan

8 Where in a central heating system should the air separator be installed?
- a Base of the open vent pipe
- b Before the circulator
- c As close as possible to the two port valve
- d On the return to the boiler

9 Which of the following systems requires a filling loop?
- a Y plan system
- b Sealed system
- c S plan system
- d Gravity system

10 Which Standard must radiators be produced to?
- a BS EN 806
- b BS EN 442
- c BS EN 12056
- d BS 1212

11 You arrive at a customer's property to find a circulator and two two-port valves located in the airing cupboard near to the hot water cylinder. What type of system could you identify?
- a Gravity heating and pumped hot water
- b Gravity heating and gravity hot water
- c Fully pumped S plan
- d Fully pumped Y plan

12 A customer asks you to advise them on a space-saving heat emitter for their small kitchen. Which one would you suggest?
- a Panel radiator
- b Fanned convector heater
- c Kick space heater
- d Low surface temperature radiator

13 In which of the following situations would it be required to install a low surface temperature radiator?
- a Nursery
- b Bathroom
- c New build
- d Conservatory

14 What is the difference between a condensing boiler and a standard boiler?
- a The flow rate to the radiators is higher
- b It can be installed in a restricted space
- c It has two heat exchangers
- d It produces instantaneous hot water

15 What is the main reason that a radiator is mounted above the skirting level height?
 a Prevent condensation
 b Allow the customer access to decorate
 c Ease of maintenance
 d Air circulation

16 You are called to a customer's property where one radiator is cold at the top and hot at the bottom when the system is heated. What action should you take?
 a Bleed the radiator
 b Drain the system
 c Turn the TRV up
 d Close the lockshield valve

17 What item takes up the expansion of the heated water in an S plan sealed system?
 a Feed and expansion cistern
 b Expansion vessel
 c The system pipework
 d Buffer in the radiators

18 As rooms in a property begin to get to temperature, the TRVs begin to close down. What valve opens up as the standard system pressure begins to increase and continues to allow the system water to flow?
 a Two port valve
 b Automatic bypass valve
 c Three port valve
 d Diverter valve

19 What would be the outward symptom in a system if the circulator was positioned after the cold feed and before the open vent?
 a Negative pressure
 b There would be no outward symptom
 c Noise in the radiators
 d Pump over

20 What is the minimum distance the open vent must rise above the feed and expansion cistern water level?
 a 150 mm
 b 250 mm
 c 350 mm
 d 450 mm

6 Sanitation and drainage (Unit 217)

You will need to know how rainwater and drainage systems work and what their component parts are along with the sanitation appliances installed.
Ask yourself:
- Why are correctly installed drainage systems so important?
- Why is trap seal loss an issue?
- What makes commissioning and maintaining systems important?

Typical mistakes

Both rainwater and drainage have been weak areas when it comes to the exam. Read through this section carefully and relate it to your workshop experience and installations at home.

Typical mistakes

Make sure you can identify the different styles of traps and how the trap seal can be lost, along with the basic installation requirements for guttering.

LO1 Understand layouts of gravity rainwater systems

Topic 1.1 Systems and materials used in gravity rainwater systems

REVISED 

The main purpose of gravity rainwater systems is to collect rainwater so that it does not:
- constitute a nuisance
- damage or ingress the building.

Rainwater runs off the roof and is collected in the gutter, which discharges the rainwater into the downpipe. This is connected to the underground drainage system.

Rainwater can be discharged into a safe outlet, such as:
- surface water drain
- combined sewer
- watercourse (stream, river, ditch)
- soakaway
- rainwater harvesting system (including a water butt).

The main types of materials used for these systems include:
- **PVCu** (most common – easy to install, long life, expands a lot)
- **extruded aluminium** (used commercially for longer length runs)
- **cast iron** (heavy, corrodes, long life, older style buildings)
- **copper** (specialist installations, turns green as it ages).

Topic 1.2 Gutter systems and components

REVISED

The different types of guttering profiles are shown in Table 6.1.

Table 6.1 Gutter shapes

Half round Standard profile used on domestic properties	
Square line Decorative style used a lot between 1980–1990s	
Ogee Modern redesign of Victorian profile, 'period' look. Sits flush with fascia	
High capacity or deep flow Deeper version of half round to cope with steeper roof lines and greater rainfall	

Check your understanding

1 Rainwater can be discharged to five locations. Can you name them?

Exam tip

Be able to identify these profiles as these shapes can come up in the exam. The shapes can give you a clue to the name.

Guttering components:

- **Running outlets** – connect the gutter to the downpipe. Forms the lowest point.
- **Gutter angles** (90° and 135°) – allow a 45° and 90° connection between lengths of gutter.
- **Gutter unions** – connect two lengths of gutter together.
- **Stop ends** – block the end of a gutter run to stop any rainwater dripping out.
- **Specialist unions** between different gutter materials – connect different materials and/or different profiles together, such as:
  - half round PVCu to Ogee PVCu
  - half round PVCu to Ogee cast iron
  - ogee PVCu to Squareline cast iron.

There are many different factors that determine the type and size of guttering systems.

Rainfall intensity:

- This is how much rain falls every second on a set area.
- It varies across the country – Cumbria has the most rain, while Essex has the least rain.
- Measured in litres per second per metre squared ($l/s/m^2$).
- BS 12056 Part 3 gives details of the installation requirements – domestic properties are Category 1.
- Where the rainfall intensity is highest (blue), deep flow profile is more likely to be used.

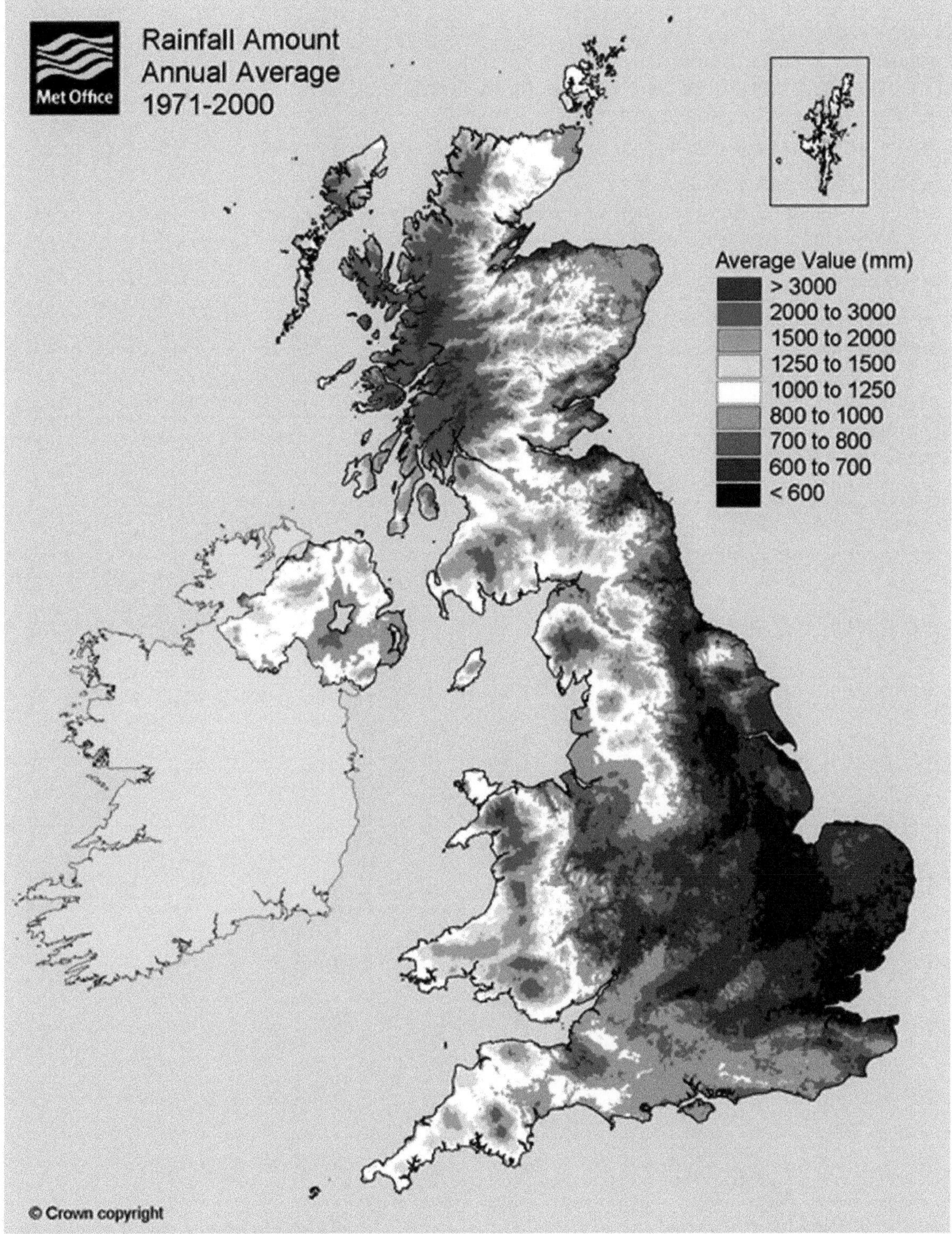

Rainfall intensity map

Roof area:

- If the roof area increases, so does the amount of rainwater collected.
- The angle of the roof is key to a gutter system. If the angle of the roof is steeper, the velocity of the rainwater running off the roof increases.
- BS 12056 Part 3 and Building Regulations Part H3 give methods to calculate the area.

Running outlet position:

- Installed above the entrance to the underground drainage system (gulley entrance).
- The more outlets, the shorter the distance the rainwater has to travel.
- Note the distances on the diagrams.

Gutter fall:

- BS 12056 Part 3 states:
  - fall between 1 mm/m and 3 mm/m
  - 1:600 ratio (for every 600 mm length the gutter drops 1 mm).
- Ensures flow capacity and cleans the gutter.
- Modern guttering can be installed level.

Change of direction in the gutter run:

- These cannot be avoided, due to the shape and roof lines of properties.
- A 90° gutter angle can restrict flow by 15%.

Customer preference:

The customer might want a particular type or size of guttering system for a specific reason, such as:

- cost
- aesthetics (the way it looks)
- age of the property
- green issues (they want to be more environmentally friendly).

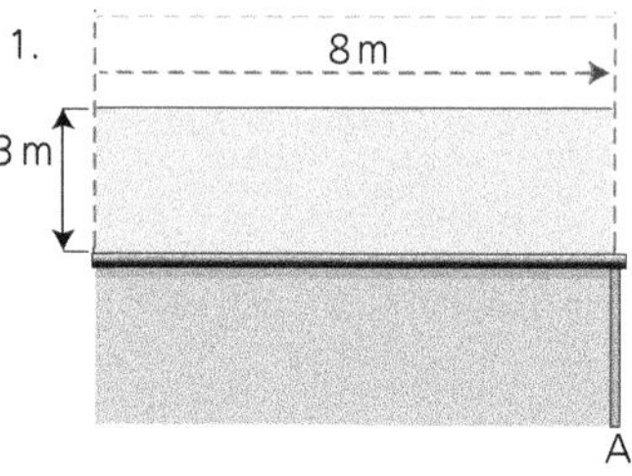

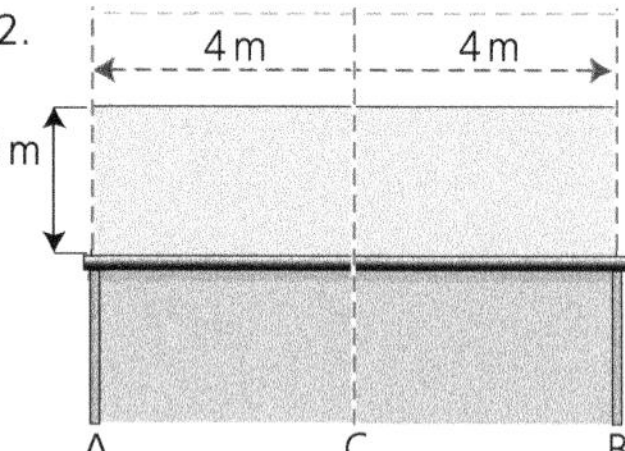

Alternative position C: Here the single outlet is equal to two outlets either end because of the outlet design

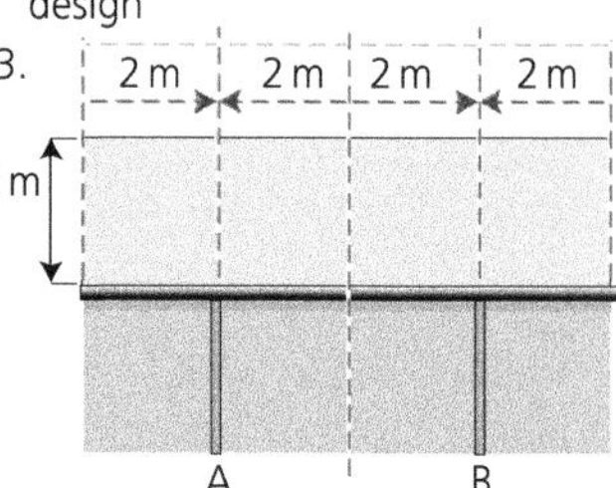

Outlet positions

Check your understanding

2 What is rainfall intensity measured in?

Now test yourself TESTED

1 You are at a customer's property which is located in a high rainfall area. What type of guttering system would you suggest they install?

2 A customer's property has a gutter measuring 3.0 m in length. Using the 1:600 ratio, work out how much fall it should have.

Topic 1.3 Rainwater pipework and components

REVISED

There are two types of rainwater pipework:

- round section (68 mm) – commonly used with the half round and deep flow profile
- square section (65 mm) – commonly used with the square line and Ogee profile.

Round downpipe

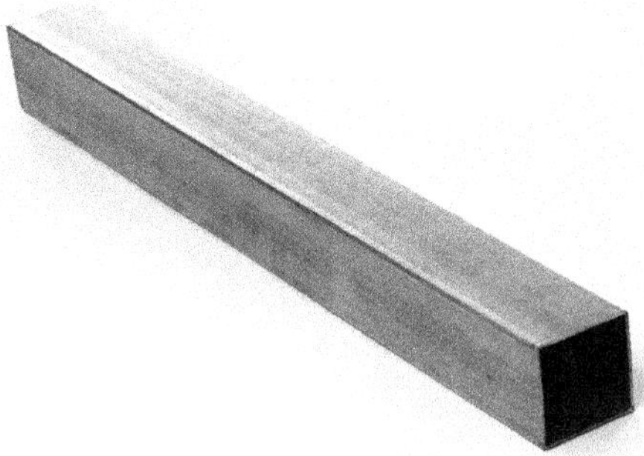

Square downpipe

The downpipe connects the gutter run to the inlet of the underground pipework.

There are also different purposes of rainwater pipework components. These include:

- **offsets** – pre-set swan neck allowing connection of two downpipes running parallel

- **angles** – 45° angle allows a swan neck to be made to any distance
- **branches** – allow the connection of a second downpipe. This is used when more than one section of gutter joins a single downpipe
- **hopper heads** – collection point for, and break in the downpipe run.
- **shoes** – bottom of a downpipe which does not directly connect to the underground drainage system
- **specialist connectors** – a drain connector joins the downpipe to the underground drainage sewer. It allows connection and seal to the underground drainage system. Generally located under a running outlet.

Check your understanding

3 What item is at the base of a downpipe and allows the rainwater to disperse away from the building?

Topic 1.4 Jointing procedures for gutter and rainwater materials

REVISED

Jointing procedures are used for gutter and rainwater materials, such as:

- **PVCu** – a snap in connection with a rubber seal
- **extruded aluminium** – silicone sealant
- **cast iron** – paint and putty; silicone sealant; rubber seal
- **copper** – silicone sealant.

Now test yourself

3 A customer has a leak in their PVCu guttering. It is between a length of guttering and a union. What two things might cause the joint to leak?

Topic 1.5 Gutter bracket selection and fixing

REVISED

This topic also includes the selection of gutter and rainwater material:

- PVCu – the most commonly used domestic material
- extruded aluminium – used where long lengths are required, more commercial
- cast iron – Victorian style, listed buildings, English Heritage situations
- copper – rarely used for domestic buildings, sometimes stately homes.

Common types of gutter bracket are described in Table 6.2.

Table 6.2 Gutter brackets

Fascia brackets	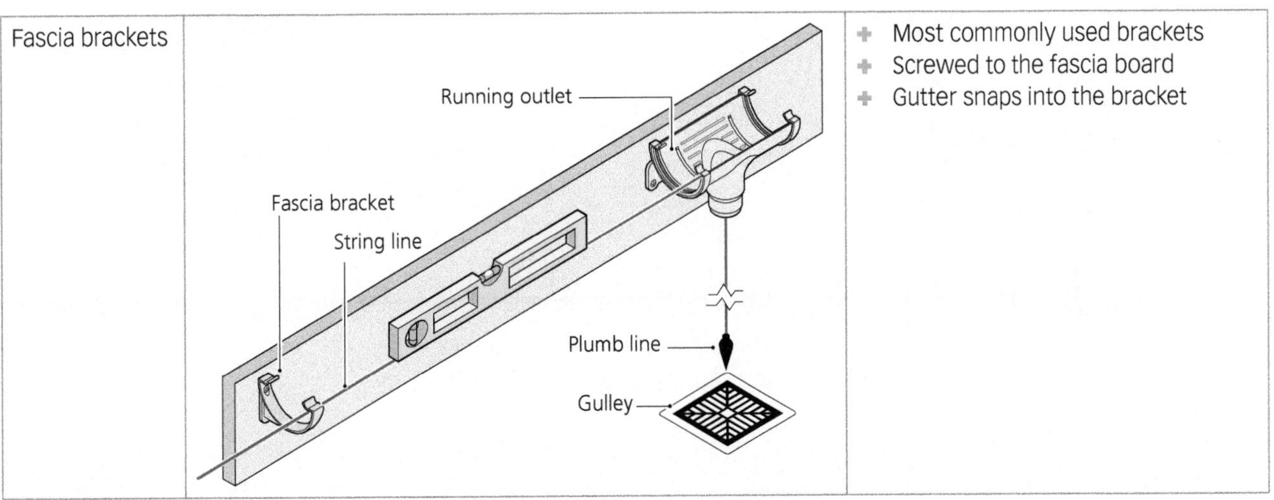	+ Most commonly used brackets + Screwed to the fascia board + Gutter snaps into the bracket

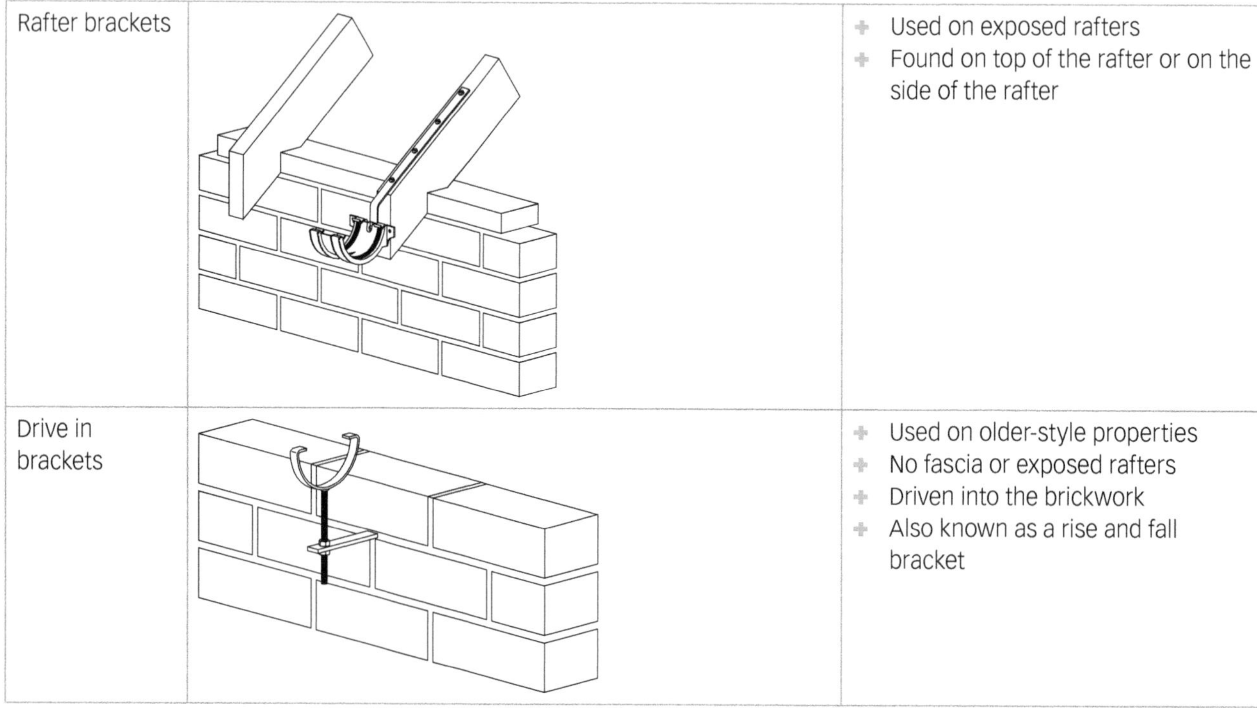

Rafter brackets		+ Used on exposed rafters + Found on top of the rafter or on the side of the rafter
Drive in brackets		+ Used on older-style properties + No fascia or exposed rafters + Driven into the brickwork + Also known as a rise and fall bracket

LO2 Install gravity rainwater systems

Topic 2.1 Sources of information for gravity rainwater systems

REVISED

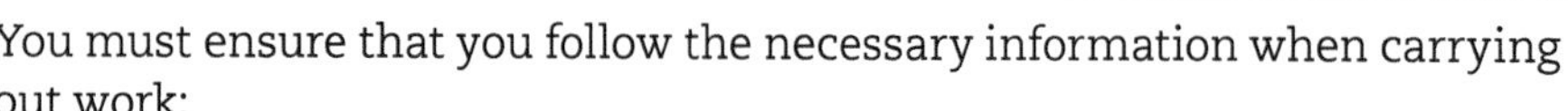

You must ensure that you follow the necessary information when carrying out work:

Statutory regulations:

- Building Regulations Approved Document Part H3.
- Design, installation and testing of gravity rainwater systems.

Industrial standards:

- BS 12056 Part 3 – design, installation, testing and maintenance.
- BS 8000 Part 13 – good workmanship.
- Other standards for pipework, fittings and material.

Manufacturers' technical instructions:

- Give details about:
  - their design for expansion
  - clipping distances
  - positioning components
  - correct fall
  - number of outlets.

Exam tip

Try to remember Building Regulations Part H covers rainwater and soil pipes. H = hole in the ground.

Topic 2.2 Preparation of the building fabric

REVISED

This is a workshop activity in which you will need to show that you know how to prepare and mount an installation depending on the building fabric. This includes:

- building wall surfaces (for example, brick, rendered, stone, stone chip, wood and cladding)
- existing gravity rainwater system components (these are covered throughout this chapter)

- incomplete building works:
  - check fascia boards are straight, level and not rotting
  - fascia boards may require painting if they are made of wood
  - check the roof underfelt has not rotted.

Always remember health and safety requirements when working. This includes:
- working at height
- risk assessments
- secure ladder; use stand-off
- inspect access equipment
- check for overhead cables
- adequate training
- being aware of your environment.

Topic 2.3 Positioning and fixing of gutter system components

This is a workshop activity in which you will need to show that you know the positioning, fixing and expansion requirements of components, including the use of the following:
- gutter brackets (fascia, rafter and drive-in types)
- running outlets
- gutter angles
- gutter unions
- stop ends
- specialist unions joining different materials or profiles.

Positioning the fascia brackets in place along the fascia board:
- Running outlet over drainage entry.
- Running outlet is the lowest point.
- 1:600 ratio fall.
- End fascia bracket highest point.
- Tight string line in between.
- Plumb line gives vertical position for downpipe.

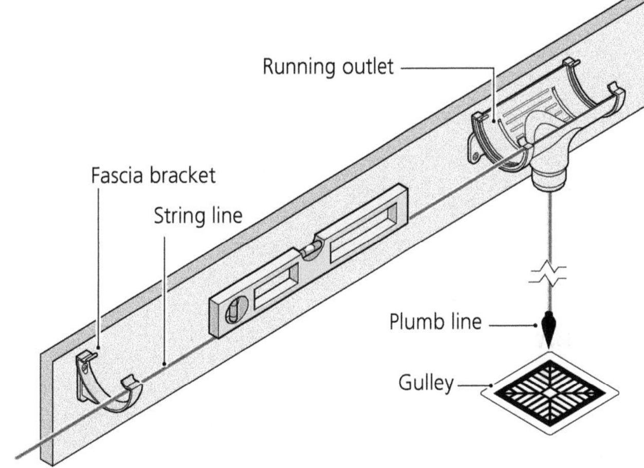

Fascia bracket position

Fixing the brackets using a string line:
- Fascia bracket max 1.0 m apart (closer if windy area).
- Use tight string line to position other fascia brackets.

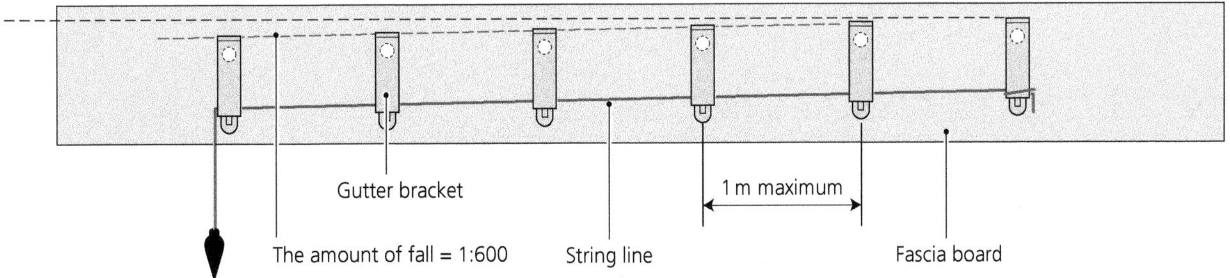

Fixing fascia brackets using a string line

Provision for the expansion of the guttering:

- Insert gutter to expansion mark not the stop.
- 10 mm expansion.
- Rubber seal.
- PVCu has coefficient of linear expansion of 0.06 mm/m/°C.

Now test yourself TESTED

4 A 5.0 m length of PVCu gutter is subjected to a 20°C temperature rise. What is the expansion?

Table 6.3 Downpipe clipping distance

Diameter of downpipe	Max horizontal	Max vertical
62 mm	1.2 m	2.0 m
68 mm	1.2 m	2.0 m

Table 6.4 Fascia bracket clipping distance

Guttering	1.0 m	Windy 0.8 m

Topic 2.4 Pipework connections

REVISED

This is a workshop activity in which you will need to show that you know the pipework connections. Take note of all connections made to the underground system. Connections can be made to:

- gulley using a shoe
- drainage bend
- direct to a gulley
- direct to a soakaway.

Topic 2.5 Install and join PVCu rainwater system components

REVISED

This is a workshop activity in which you will need to show that you know how to install and join PVCu rainwater system components. Take note of all the component parts that are used, along with their function, and have another look at LO1 (Topics 1.2 and 1.3). The components include:

- running outlet
- gutter angle
- gutter union
- stop end
- downpipe
- brackets.

Topic 2.6 Test rainwater systems

REVISED

- This is a workshop activity in which you will need to carry out tests on gravity rainwater systems, including suitable operation and leakage.
- Testing is carried out by discharging some water from a hosepipe onto the roof line towards the high points of the system.
- Check the water enters the gutter correctly, flows down towards the outlet, and that there is no pooling or any leaks.

LO3 Understand service, maintenance requirements and commissioning of gravity rainwater systems

Topic 3.1 Maintenance checks

REVISED

Maintenance checks are used to ensure gravity rainwater systems are working correctly and safely. You need to know how to carry out a visual inspection as part of a routine check and how to correct defects in systems.

Adequate support:

- This is to establish the overall condition. Check that:
  - all clips are intact
  - there are sufficient clips
  - there are not any low dips
  - fall is in the correct direction.
- Replace any broken clips or add more clips if required. Correct the fall direction if necessary.

Leakage and damage:

- This can be determined if there are any of the following features:
  - damp marks on the building
  - moss build-up
  - rust marks
  - broken or cracked parts.
- You must inspect the seals and replace any rubber seals or components if required.

Obstructions:

- This includes leaf and debris build up. Check that the outlet entrance is clear and remove any obstructions. Fit a leaf guard if required.

Now test yourself

TESTED

5 A customer complains that their guttering pools with rainwater and does not drain properly. What could be the cause of this pooling?

Topic 3.2 Defects in systems

REVISED

Defects that you will come across include:

- leaks – due to cracks or damaged seals
- blockages – due to falling leaves, contaminated waste or obstacles
- support – inadequate or broken brackets.

Take care with working at heights.

The use of a 'stand-off' attached to the ladder may be required.

Gloves or gauntlets will be required.

Now test yourself

TESTED

6 You are called to a customer's property where the guttering system overflows every time it rains. What would be the primary cause of this overflow?

Topic 3.3 Pre-commissioning checks

REVISED

The types of checks to be carried out before commissioning are:

- check against the installation requirements – drawings, plans, job specification, customer's requirements
- brackets and support – 1.0 m distance, correct number, no broken brackets
- jointing – seals are in place and secure. Allow for expansion
- remedial work – complete any additional work to bring the installation up to standard.

LO4 Understand sanitary appliances

Topic 4.1 Working principles of sanitary appliances

REVISED

Below are the types and working principles of different sanitary appliances.

> **Exam tip**
>
> Look at the shape of the WC pans and be ready to identify each style.

Conventional WC (toilet)

Wash down:

- Most common type.
- Force of water flushes.

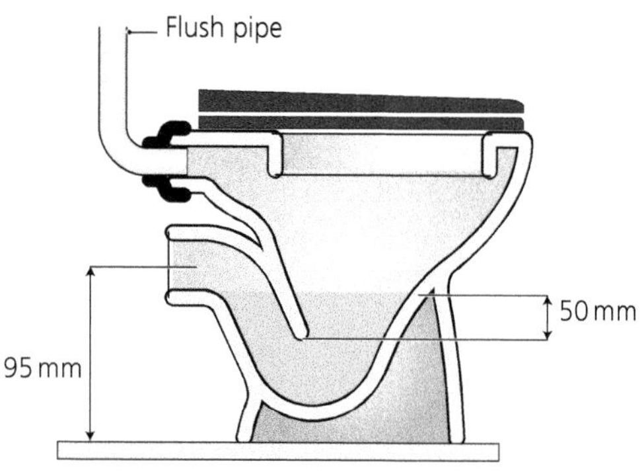

Wash down

Single trap siphonic:

- Restrictions create a vacuum, clearing the flush.

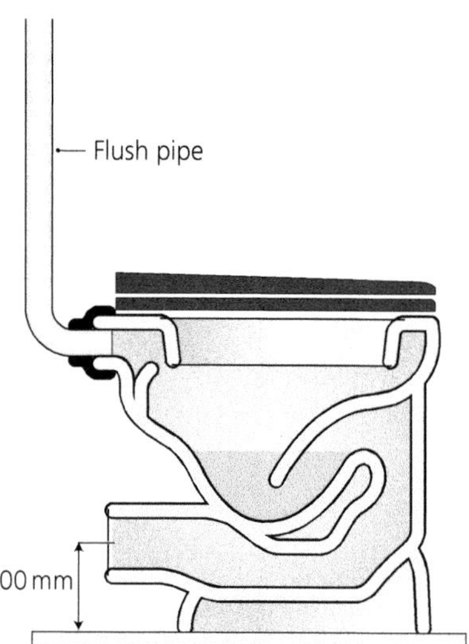

Single siphonic

Double trap siphonic:

- Rarer type.
- Aspirator creates negative pressure, which clears the flush.

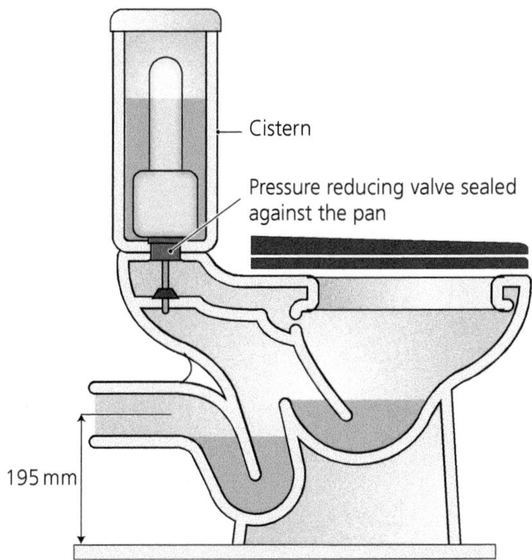

Double siphonic

Different styles:

- **Close coupled** – the cistern is connected directly to the WC pan via a rubber 'doughnut' to keep a watertight seal.
- **Low level** – the cistern is 1 m above ground level, with a 32 mm flush pipe to the WC pan.
- **High level** – the cistern is 2 m above ground level, with a 40 mm flush pipe to the WC pan. Uses a pull chain to flush (for example, old outside WC).
- **Back to wall** – cistern is concealed behind a panel. Pan sits on the floor and flush to the wall.
- **Wall hung** – cistern is concealed behind a panel. Pan is bolted to a framework behind the panel, which allows cleaning under the pan.

When flushing a dual flush WC, a part flush is 4.5 litres and a full flush is 6 litres.

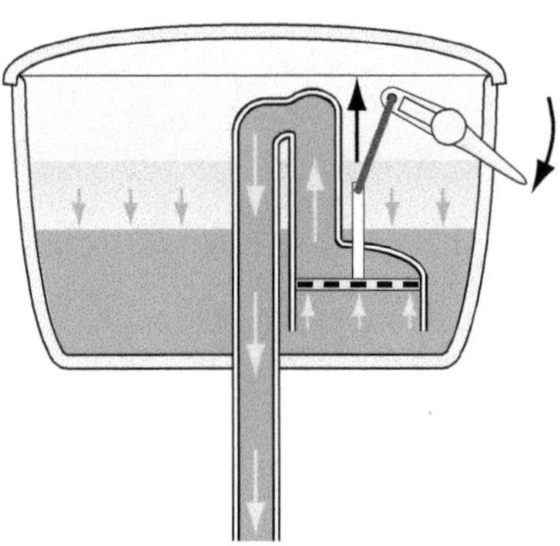

A siphon uses siphonic action to flush.

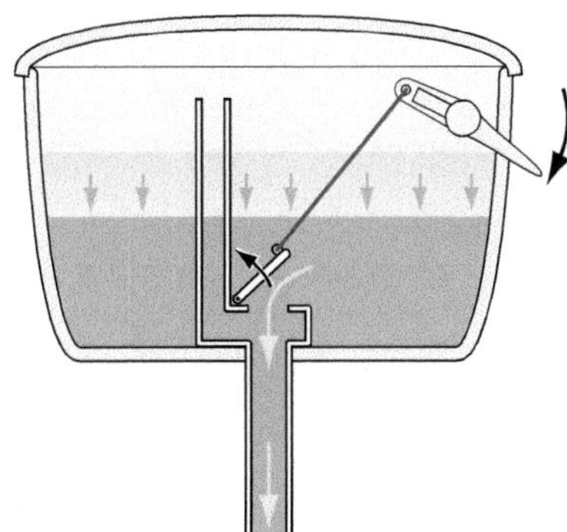

A flush valve uses gravity to flush.

When the handle is depressed, the flap lifts, allowing water to flow to the pan by gravity

Check your understanding

4 What is the name for a WC (toilet) that allows the customer both a full flush and a part flush?

Baths

The common bath sizes are 1600–1800 mm long × 700–800 mm wide. Baths use a combined waste and overflow waste fitting, connecting the overflow to the waste via a flexible tube.

Different types of bath:

- **Standard bath** – standard bath with bath panels.
- **Corner bath** – installed in the corner, uses a curved bath panel. They tend to use a lot of water.
- **Off-set corner bath** – sides are unequal to utilise space. Left-hand and right-hand versions are available.
- **Free standing bath** – stand on their own exposed feet, no supporting wall. Roll top with claw feet.
- **Double-ended bath** – two ended with taps in the middle.
- **Tapered bath** – used where space is limited, can have shower at wider end.
- **Shower/bath** – 'P' shape bath to allow space for shower.
- **Whirlpool or spa bath** – pumped air and water. Take care when fitting as it is connected to electricity.

Bidets

Different types of bidet:

- **Over the rim** – provides an air gap to prevent contamination.
- **Ascending spray** – no air gap so risk of contamination. It cannot be connected to the mains or combi boiler, supply via cistern.

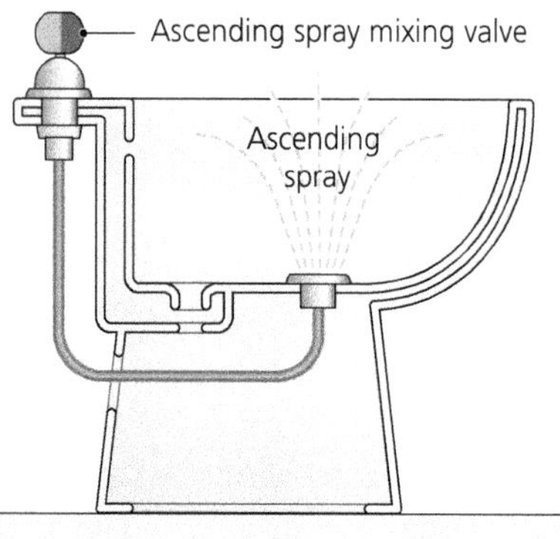

Ascending spray bidet

Now test yourself TESTED

7 At a customer's property you are asked to install an ascending spray bidet. On inspection, you observe the property has a combination boiler. Outline what you will explain to the customer.

Wash hand basins

Basins use a slotted waste for the overflow and allow a plug. Alternative is a pop-up waste.

Different types of basin:

- **Wall hung** – mounted on the wall using fixing brackets or bolts (no legs). The mounting must be able to take the weight of the unit.
- **Pedestal** – fixed to the wall but it rests on a pedestal mounted on the floor. The pedestal is designed to hide the pipework.
- **Semi-pedestal** – fixed to the wall but rests on a pedestal mounted on the wall below the basin.

- **Countertop** – fitted to vanity units and countertops:
  - countertop basin or inset – whole basin is inset into the unit and rests on the rim
  - semi-countertop basin – part of the basin is inset into the unit and part overhangs
  - under-countertop – whole basin is mounted under the work surface.
- **Vessel** – rests completely on top of the work surface.

Different types of taps:

- **One hole** – mixer style tap.
- **Two hole** – independent hot and cold tap.
- **Three hole** – independent spout with two separate wheel heads for the hot and cold.
- **No hole** – taps are wall or unit mounted separate to the basin.

Shower trays and cubicles

- No overflow.
- Use a flush mounted removable waste (removeable for cleaning).
- Raised lip to the tray leading to waste.
- Cubicles installed to suit the setting.
- The cubicle design varies a lot from: door only; side and door; and free standing.

Sinks

Sinks need to be robust and hardwearing.

Different styles of sink:

- Standard kitchen sink – single bowl/single drainer, double bowl/single drainer, vegetable trough.
- Butler sink – like a London sink, but has a high splash back.
- Cleaner's sink – low level heavy sink on legs. Allows buckets to be filled.
- Belfast sink – heavy sink with weir style overflow. Retro fit in kitchens.
- London sink – same as a Belfast sink, but no overflow.

Urinals

- **Bowl type.**

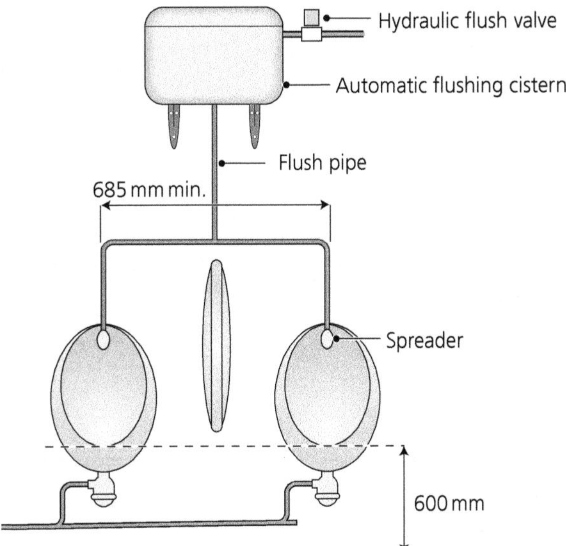

Bowl type urinal

- **Trough type** – tend to be wall mounted stainless steel.
- **Slab type.**

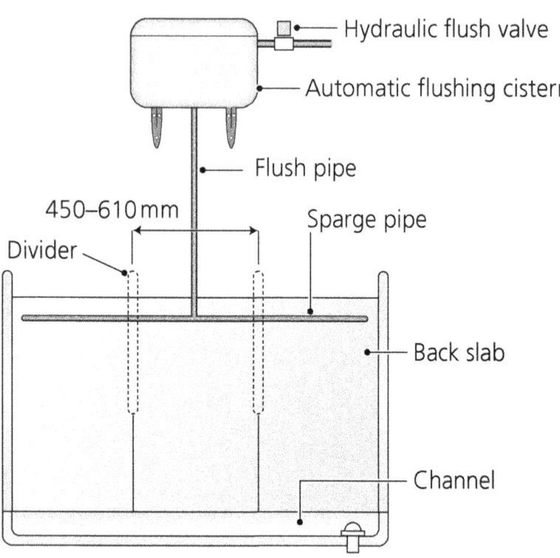

Slab type urinal

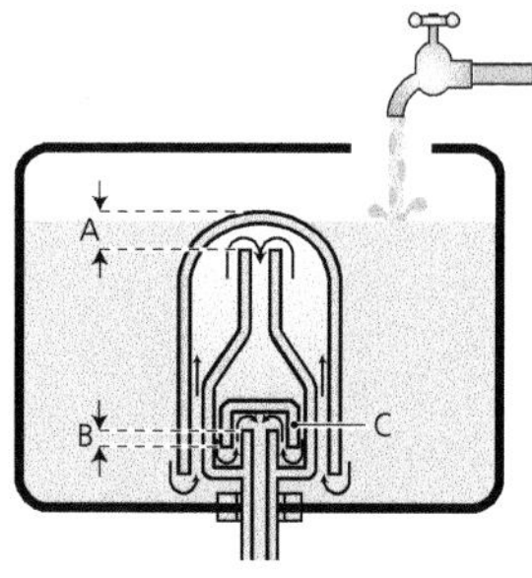

Automatic flush valve

When flushing, it takes 10 litres per hour for a single bowl and 7.5 litres per hour for a multi-position.

An automatic flush valve/automatic flushing siphon:

+ works on differential pressure
+ the pressure at point A becomes greater than at point B as the cistern fills.

Check your understanding

5 According to the Water Supply (Water Fittings) Regulations, which bidet offers the greatest hazard of water contamination?

Topic 4.2 Features of sanitary pipework and layout

REVISED

Types of soil stack

Table 6.5 Types of soil stack

Primary ventilated stack system + Most common domestic style + Relies on atmospheric pressure being maintained. What goes down as waste, must come up as air movement + Top section is the vent + Lower section is the soil stack	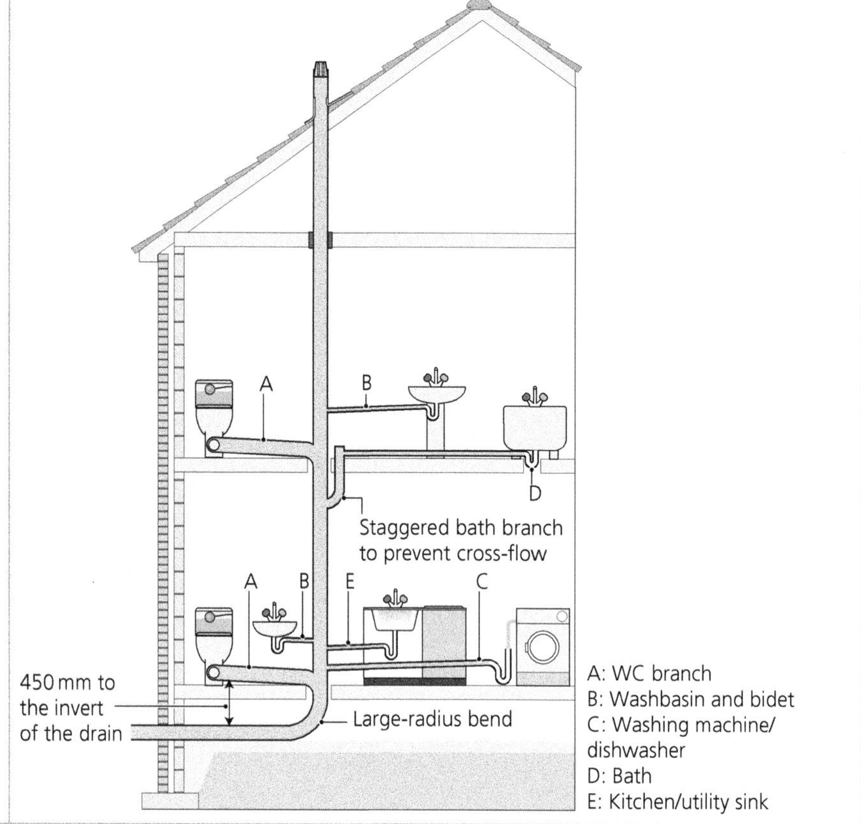

Secondary ventilated stack system + More complex and a larger installation + Second vent pipe allowing additional air movement for large discharge volumes	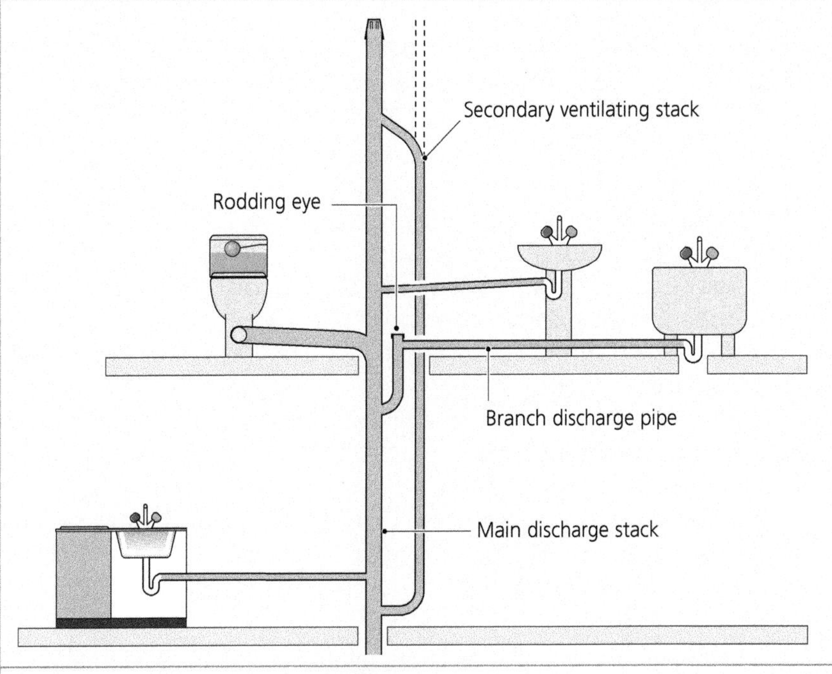
Ventilated branch discharge system + Most complex for very large installations + Secondary vent pipe and individual vent pipes to each branch	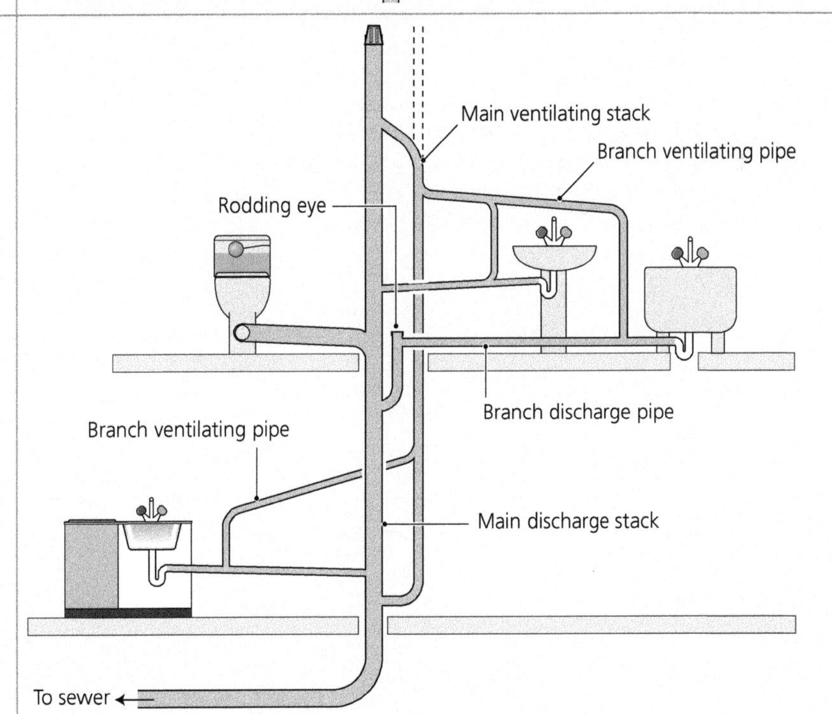

Branch and waste sizes

Table 6.6 Branch and waste sizes

Appliance		Pipe size (mm)	Max. length (m)	Gradient (mm/m)	Trap seal depth (mm)
A	WC branch	75–100	6	18	50
B	Washbasin and bidet	32	1.7	18–22	75
C	Washing machine/ dishwasher	40	3	18–90	75
D	Bath	40	3	18–90	50
E	Kitchen/utility sink	40	3	18–90	75
Where these lengths are exceeded, then the next pipe size up should be used; 40 mm appliances will need to increase to 50 mm pipe, the length and gradient of which are listed below.					
Appliances with 50 mm waste pipe			4	18–90	75

Check your understanding

6 When installing a basin, what is the recommended fall for the waste pipe?

Key factors

- At the base of the soil stack, there must be a long radius bend.
- Standard soil pipe diameter is 110 mm.
- Must prevent cross-flow by allowing a 200 mm distance between a WC connection and any branch pipe opposite.
- If the soil stack terminates inside the roof line, an air admittance valve should be fitted.
- Branch pipe size must never get smaller, as this will restrict the flow of water and cause trap seal loss.
- Multiple appliances connected via 50 mm waste pipe.
- If the soil stack terminates within 3 m of an opening window, the stack must continue to rise 900 mm above the window.
- The cage on top of the soil stack prevents birds and debris falling down inside.
- Purpose-made access needs to be installed at the base of the soil stack.

Soil and waste pipe clipping distances

Table 6.7 Soil and waste pipe clipping distances

Diameter of soil/ waste pipe	Max. horizontal	Max. vertical
32 mm	0.5 m	1.2 m
40 mm	0.5 m	1.2 m
50 mm	0.9 m	1.2 m
110 mm	1.0 m	2.0 m

Topic 4.3 Ground floor systems and appliances

REVISED

The layout features and connections for ground floor systems and appliances are shown below.

Stub stack systems:

- Ventilation is required if the stub stack is further than 6 m from the main stack, or the highest connection is above 2 m.
- WC connection heights – max. 1.5 m from the invert of the drain.

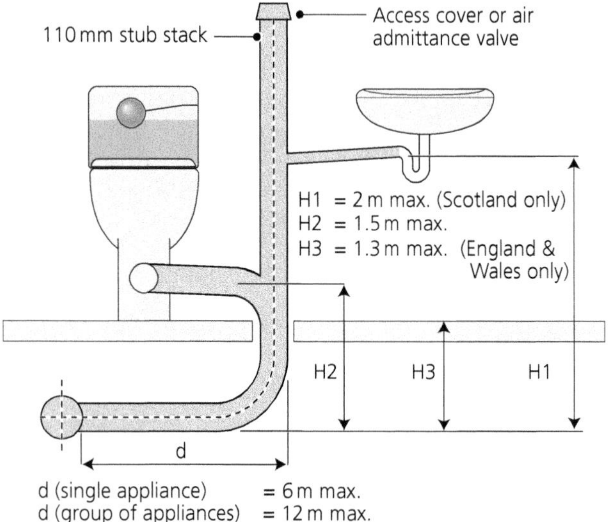

Stub stack

Connections to gullies:

- Waste pipes may discharge into a gulley if:
  - the gulley is capable of accepting the discharge
  - the waste pipe is below the gulley grate
  - the appliance using the gulley has a 38 mm trap.

Connections direct to drain:

- Connections may vary according to the age of the underground system.
- Salt glaze and earthenware via a collar, which is sealed by sand and cement.
- PVCu uses a push fit sleeve into the long radius bend.

WC connection direct to drain:

- Lowest connection for a low-rise building – 450 mm.
- Lowest connection for a high-rise building – 750 mm.

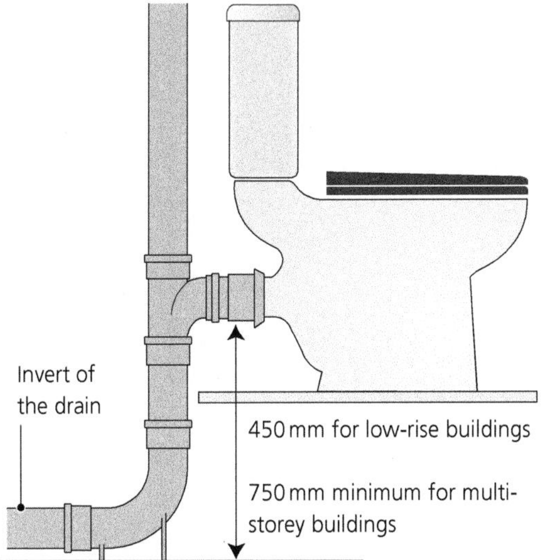

WC to drain

Check your understanding

7 What is the maximum distance a stub stack can be installed from a ventilated stack?

8 Why must a long radius bend be installed at the base of a soil stack?

Exam tip

Questions in the exam may use the term 'invert'. The invert is the centre line of the underground drain which will fall (get deeper) towards the road.

Topic 4.4 Types of traps and seal loss

Different types of traps

Trap seal

- The purpose of a trap seal is to prevent the noxious (nasty) smell entering the property.
- Use compression fittings to connect to the appliance and the discharge branch.
- Watertight seal via a rubber seal.

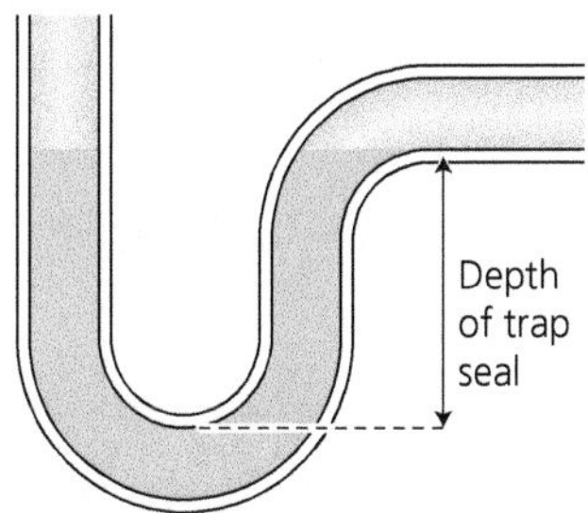

Trap seal depth

Waste pipe size and trap seal depth:

- The 'waste' is installed in the appliance and connects the appliance to the trap.
- The more water to discharge, the larger the waste pipe diameter.
- If the appliance has no overflow, or it is a pumped discharge, the waste pipe is also larger.

Table 6.8 Waste pipe size and trap seal depth

Appliance	Waste fitting size (inches)	Diameter of trap (mm)	Trap seal depth when fitted to a primary ventilated system (BS EN 12056-2) (mm)
Washbasin	1 ¼	32	75
Bidet			
Bath	1 ½	40	50
Shower			
Bowl urinal		40	75
Washing machine		40	75
Dishwasher			
WC pan	N/A	75	50
		100	50

Swivel P-trap:

- Often used on new and replacement appliances.
- 32 and 40 mm.
- Under kitchen sinks.
- Discharge branch at 90°.
- Available with a spigot trap, which allows a washing machine hose to be connected.

Swivel S-trap:

- Often used on new and replacement appliances.
- 32 and 40 mm.
- Under basins.
- Prone to capillary action trap seal loss.
- Discharge vertical.

In-line trap (slimline trap):

- S trap style designed for pedestal basins.
- Prone to capillary action trap seal loss.

Washing machine trap:

- P trap with 600 mm height.
- Open top to allow washing machine hose to enter.

Bottle trap:

- Used on basins and bidets.
- Access to clean via base.
- If used on urinals they are prone to blocking.
- Must never be used under kitchen sinks – can cause blockages.

Bath trap:

- Specially designed for confined spaces with either a 50 mm or 38 mm trap seal.

Shower trap:

- Bottle-type trap with access for cleaning from the top or shower tray.

Anti-vac trap:

- No substitute for a well-planned and designed system. If designed to BS 12056 Part 3, there should be no trap seal loss.
- Uses a small air admittance valve to allow additional air into the system to avoid negative (sucking) pressure.
- Can be fitted to any style of trap.

Waterless valve:

- Not officially a trap.
- Neoprene rubber sleeve opens under the pressure of water passing through and closes after all the water has finished flowing.

Exam tip

If a question asks you to identify an image of a trap, look carefully in case it is an anti-vac trap with the small air admittance valve attached. The answers offered could include both trap names (P-trap or anti-vac P trap).

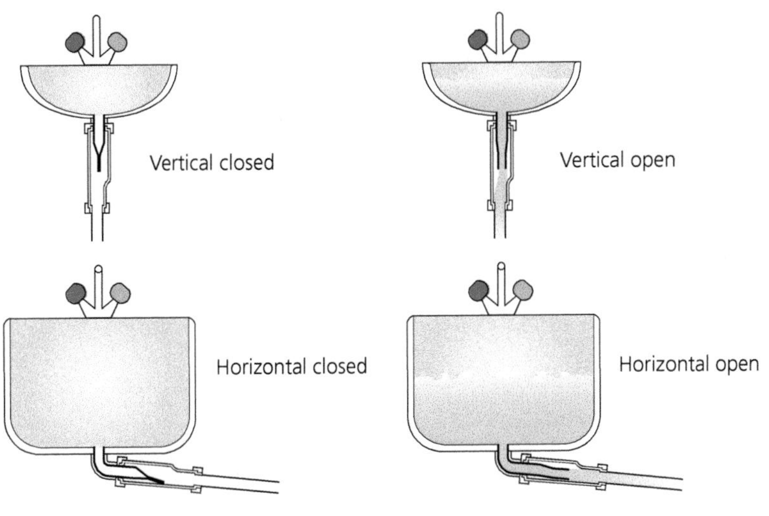

Waterless seal

> **Check your understanding**
>
> 9 Why has a shower trap got a removable grill?

Factors that lead to trap seal loss

Table 6.9 Factors that lead to trap seal loss

Incorrect installation	If the system is not designed to BS 12056 Part 3, then trap seal loss could be a consequence	+ Waste pipe is too long or too small + Incorrect fall + No long radius bend + Too many appliances on one branch + Too many changes in direction
Wavering out	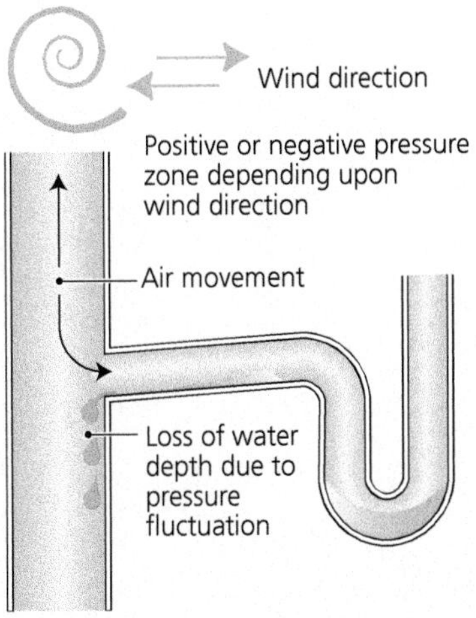 Wind direction Positive or negative pressure zone depending upon wind direction Air movement Loss of water depth due to pressure fluctuation	+ Caused by wind passing over the top of the stack, changing the air pressure in the top of the stack + This 'rocks' the top traps and washes a small amount away each time + Install a cowl on top of the stack to prevent it
Induced siphonage	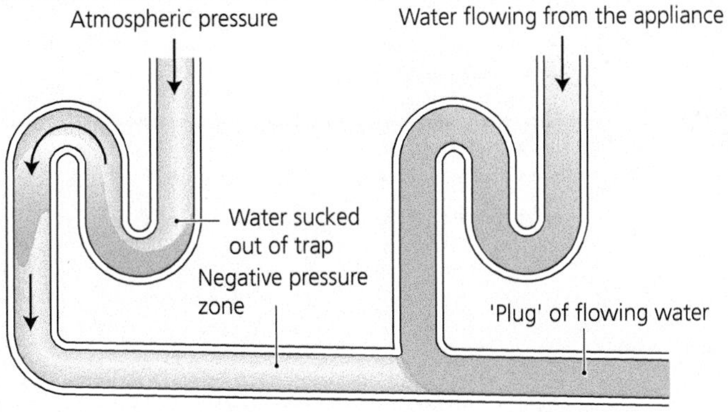 Atmospheric pressure Water flowing from the appliance Water sucked out of trap Negative pressure zone 'Plug' of flowing water	+ One appliance pulls another appliance trap out + An area of negative pressure is formed, pulling the trap out

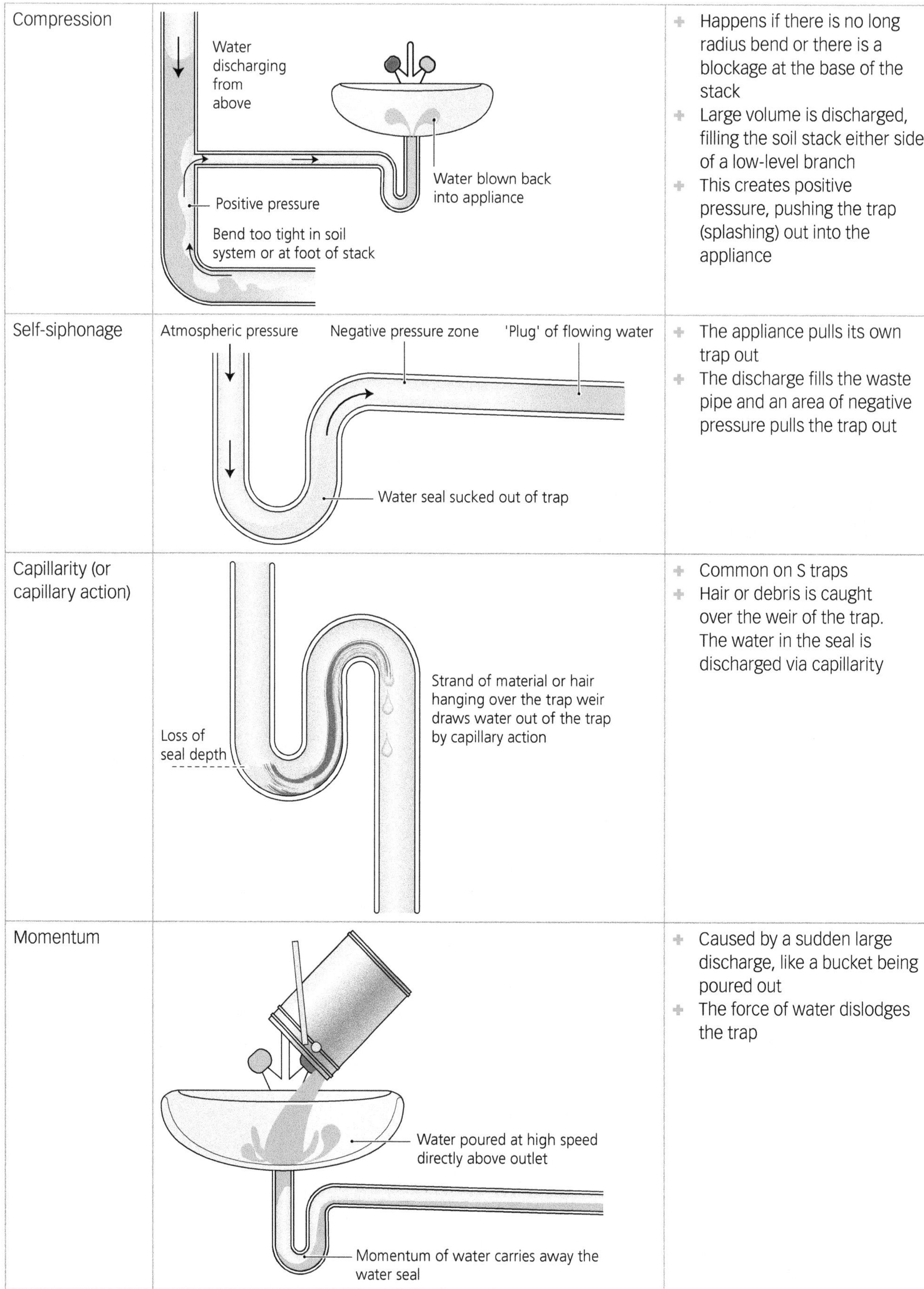

Compression	Happens if there is no long radius bend or there is a blockage at the base of the stack Large volume is discharged, filling the soil stack either side of a low-level branch This creates positive pressure, pushing the trap (splashing) out into the appliance
Self-siphonage	The appliance pulls its own trap out The discharge fills the waste pipe and an area of negative pressure pulls the trap out
Capillarity (or capillary action)	Common on S traps Hair or debris is caught over the weir of the trap. The water in the seal is discharged via capillarity
Momentum	Caused by a sudden large discharge, like a bucket being poured out The force of water dislodges the trap

Evaporation	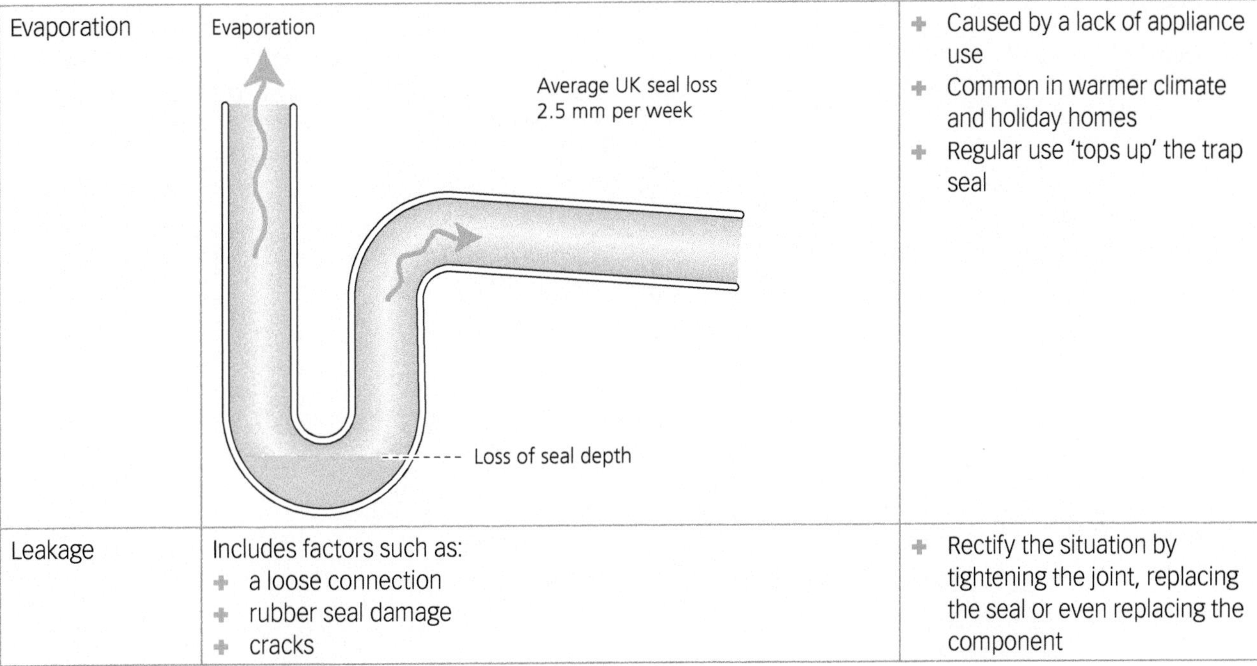	+ Caused by a lack of appliance use + Common in warmer climate and holiday homes + Regular use 'tops up' the trap seal
Leakage	Includes factors such as: + a loose connection + rubber seal damage + cracks	+ Rectify the situation by tightening the joint, replacing the seal or even replacing the component

Check your understanding

10 If an appliance suffered from continual self-siphonage, what would be the most convenient way of rectifying the situation without altering the pipework?

Exam tip

Capillary action is the same as capillarity, which is sometimes used in questions.

Topic 4.5 Drainage systems

REVISED

Different drainage systems are required depending on their suitability to receive foul soil and waste water.

Table 6.10 Below ground drainage systems

Combined	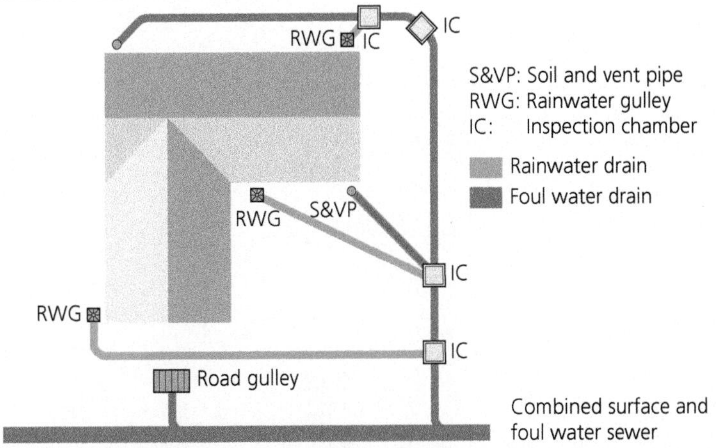 RWG, IC, IC, S&VP, RWG, IC, RWG, IC, Road gulley S&VP: Soil and vent pipe RWG: Rainwater gulley IC: Inspection chamber Rainwater drain Foul water drain Combined surface and foul water sewer	+ Both the foul and rainwater discharge into a common sewer + It is an older system + Rainwater flushes foul sewer out + It is cheaper as there is less pipework + Impossible to connect foul water to the wrong drain as all pipes lead to the foul sewer + All discharge goes to water treatment plant, which is expensive
Separate	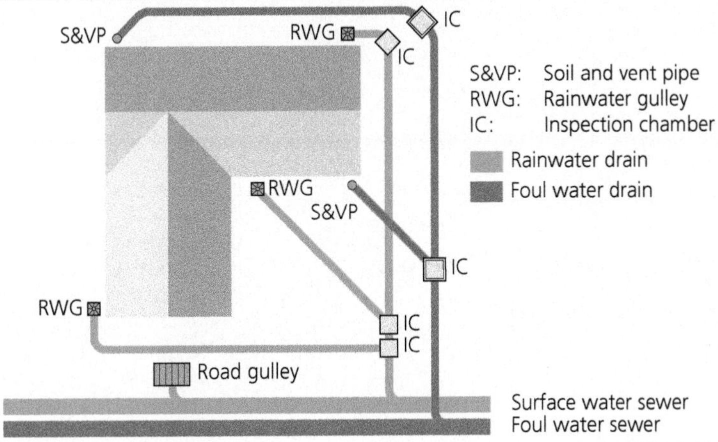 S&VP, RWG, IC, IC, RWG, S&VP, IC, RWG, IC, IC, Road gulley S&VP: Soil and vent pipe RWG: Rainwater gulley IC: Inspection chamber Rainwater drain Foul water drain Surface water sewer Foul water sewer	+ Favoured by local authorities + Separate pipework for the foul water, as well as for the rainwater + Expensive to install as there is a lot of pipework + Care needed when connections made so foul and rainwater do not cross over + Water treatment is cheaper as the discharges are separate

Partially separate	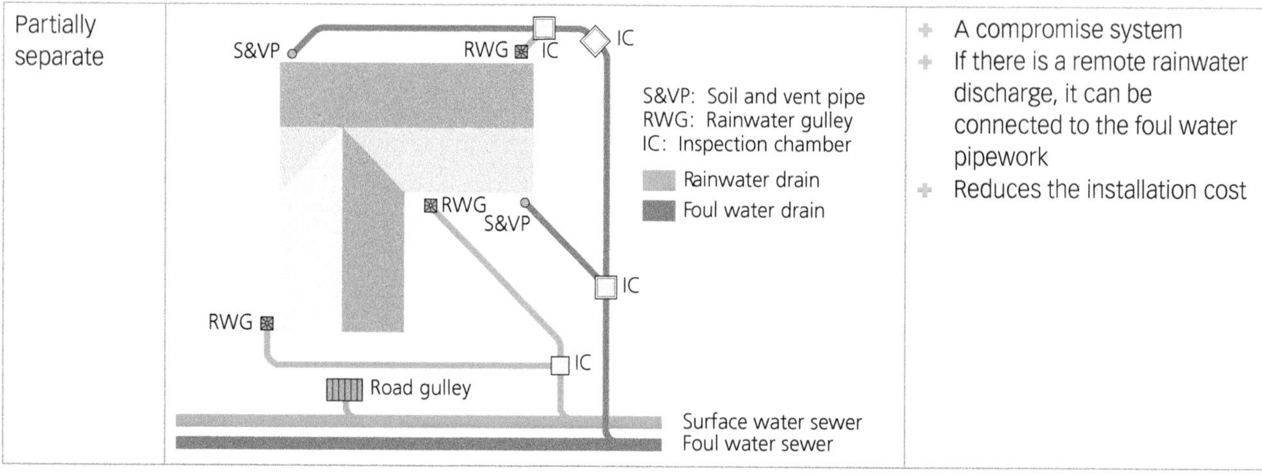	+ A compromise system + If there is a remote rainwater discharge, it can be connected to the foul water pipework + Reduces the installation cost

Check your understanding

11 On a separate system, what should you do if a connection for a WC has been made onto the rainwater pipework?

Topic 4.6 Condensate drain connections

Condensate drain connections are made below every condensing boiler, where internally in the boiler, the condensate is formed. It is important to know the correct termination for the condensate as it is slightly acidic – it must be disposed of carefully, as outlined in the manufacturer's instructions.

Connections to a trap:

+ Most condensing boilers have a 75 mm internal trap, unless the manufacturer states otherwise.
+ Must have a trap otherwise the gas way has been broken and this contravenes the Gas Regulations. This would potentially allow flue gases to escape.

Connections to a drain:

+ Must not connect to rainwater system as the condensate is slightly acidic, so it must enter the foul water system.

Pipe sizes and insulation:

+ Internal – 21.5 mm overflow pipe.
+ External – 21.5 mm overflow pipe fully insulated. 32 mm waste pipe externally with no insulation.
+ Pipe must be PVCu, ABS, PP or PVC.
+ No metal pipework as condensate is acidic.

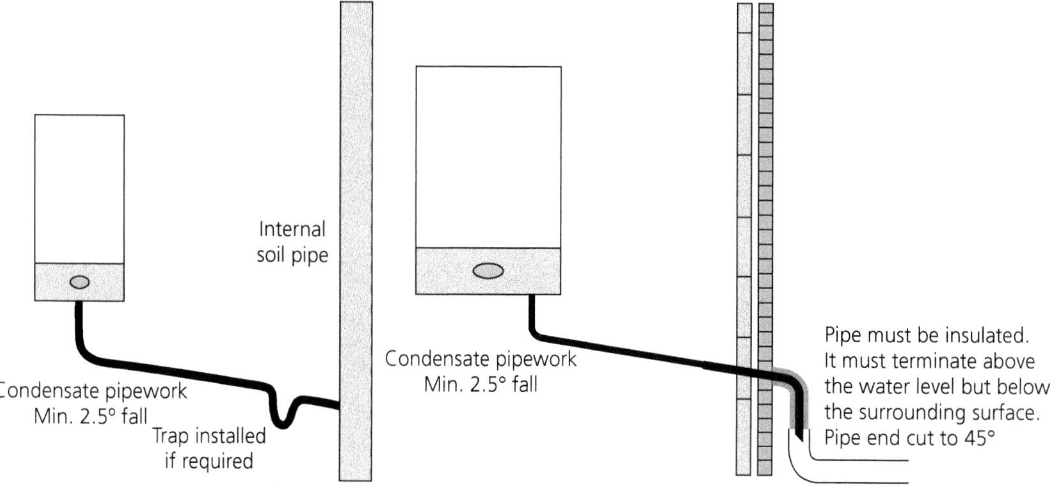

Condensate pipe

Gradient:

- Fall must be at least 2.5°, or 50 mm every metre.
- The length of pipe should not be over 3.0 m.
- Outside pipework should be as short as possible.

LO5 Install sanitary appliances

Topic 5.1 Sources of information; Topic 5.2 Installation requirements of appliances and systems; Topic 5.3 Decommissioning process of appliances and systems

This combines with a workshop activity in which you will need to show that you know the methods used to install sanitary appliances and connecting pipework systems, including the use of the following:

- BS 12056 Part 2 and Part 5 – Gravity drainage, installation, testing and operation
- BS 8000 Part 13 – Good workmanship
- knowing the correct information to be referred to when carrying out work (statutory regulations, industry standards, manufacturers' technical instructions and design requirements)
- preparing work required for installation (storage requirements, assembly of appliances and preparing building fabric)
- knowing what information should be provided to other users before decommissioning (e.g. the length of time the system may be out of use)
- methods used to reduce periods when facilities are not available (work on a section at a time, work outside office hours)
- methods used to prevent the end user from using the appliance or system (temporary capping, warning signs)
- permanent and temporary decommissioning processes.

Topic 5.4 Install and test systems and appliances

You must know how to carry out installation of primary ventilated stack systems and appliances, as well as how to carry out tests before the operation, such as visual inspections and air tests. These tasks will be combined in workshop activities for you to undertake. Once you have installed the system, you will have to visually inspect the system and carry out a soundness test on the system.

Table 6.11 Post installation inspection

Visual inspection	+ Visual inspection involves checking:
	  + connections
	  + gradients
	  + clips
	  + termination
	+ Set appliances:
	  + water levels
	  + mechanisms
	+ Installed according to BS 12056 Part 3	

Air test (or soundness and performance)	Ensure that there are no leaks and the system performs to BS 12056 Part 3 **Soundness test:** + Seal ends and fill traps + Hand pump to 38 mm + Hold for 3 minutes Larger systems may have to be tested in sections. **Performance test:** + Fill appliances with water to overflow level + Empty/Flush all appliances at the same time (simulating worst conditions) + Inspect each trap seal depth with a matt black dip stick + Minimum of 25 mm trap seal must be left + Carry out three times	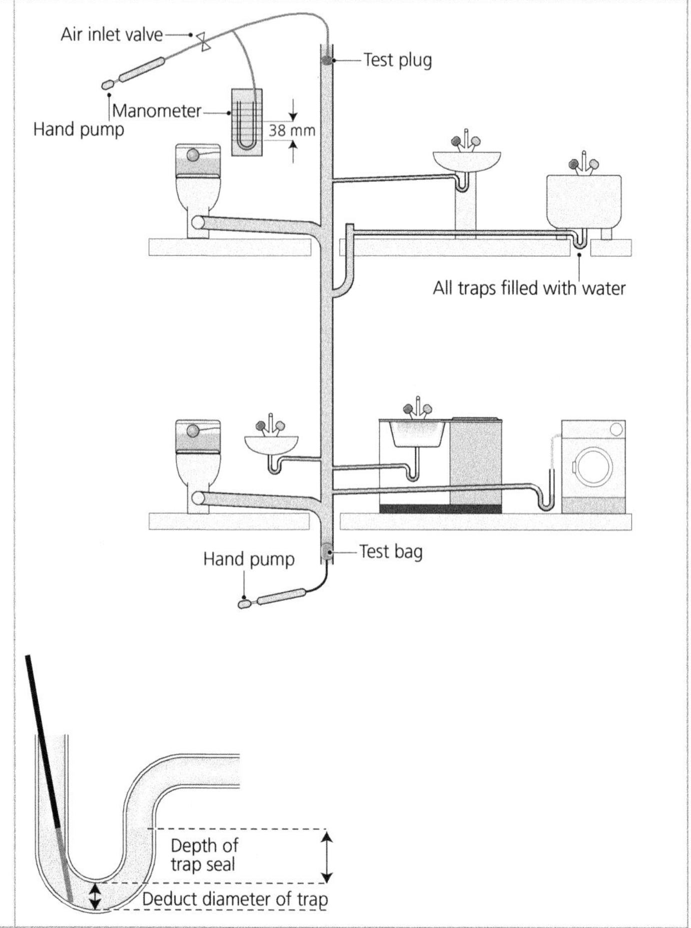

Check your understanding

12 At what pressure is a soundness test on a soil stack carried out?

LO6 Understand service and maintenance requirements

Topic 6.1 Maintenance checks

REVISED

Visual inspections will need to be performed as part of routine checks. This will involve:

+ Cleaning out traps.
  + They accumulate debris and may smell.
+ Cleaning out overflows of appliances.
  + Notorious for blocking.
+ Checking access covers.
  + Check for leaks.
  + Lubricate screws and bolts.
+ Inspecting pipework. Look out for:
  + leaks
  + corrosion
  + broken clips.
+ Inspecting water levels.
  + In WC cisterns.
+ Checking for slow discharges.
  + Suggests there is a blockage.

Topic 6.2 Defects in systems

REVISED

When correcting defects, you will need to perform visual inspections, use appropriate equipment and wear the correct PPE.

PPE:

- eye protection
- face protection
- rubber gauntlets
- body protection.

PPE can help to protect you against hepatitis (caught from foul water) and Weil's disease (from rat's urine).

Tools used to clear blockages:

- **Drain rods** – remove major blockages in a stack or underground system.
- **Force cup** – clears a basin or sink blockage.
- **WC plunger** – moves a greater volume of air than a force cup, so only used for WCs.
- **Spinner** or **auger** – long spring fed through a waste into the branch. Pipe which rotates and releases debris in the waste pipe of a basin, sink or bath.

Using chemicals:

- Take great care when using chemicals and follow the manufacturer's instructions.
- Do not mix chemicals.
- Flush away completely after use.

Exam-style questions

1 Identify this guttering profile:

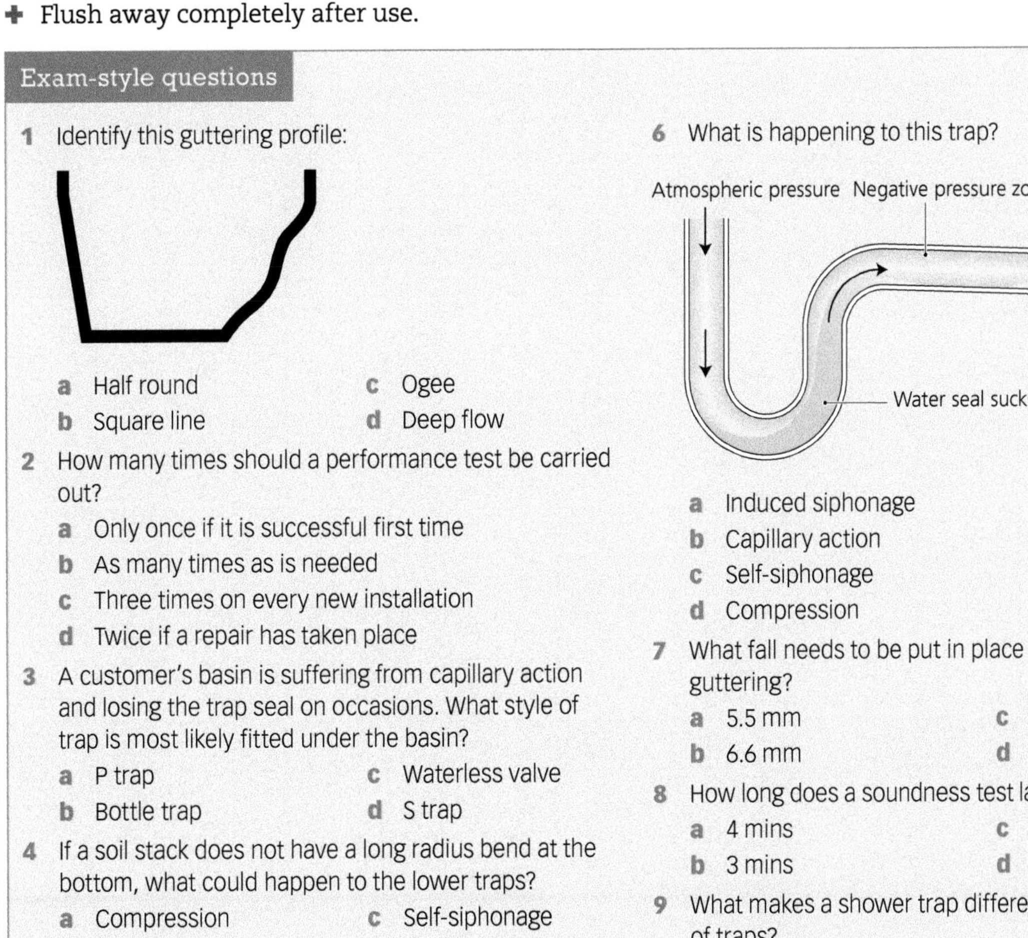

a Half round
b Square line
c Ogee
d Deep flow

2 How many times should a performance test be carried out?

a Only once if it is successful first time
b As many times as is needed
c Three times on every new installation
d Twice if a repair has taken place

3 A customer's basin is suffering from capillary action and losing the trap seal on occasions. What style of trap is most likely fitted under the basin?

a P trap
b Bottle trap
c Waterless valve
d S trap

4 If a soil stack does not have a long radius bend at the bottom, what could happen to the lower traps?

a Compression
b Wavering out
c Self-siphonage
d Capillary action

5 What is the maximum clipping distance for fascia brackets on a guttering system?

a 700 mm
b 800 mm
c 900 mm
d 1000 mm

6 What is happening to this trap?

Atmospheric pressure
Negative pressure zone
'Plug' of flowing water
Water seal sucked out of trap

a Induced siphonage
b Capillary action
c Self-siphonage
d Compression

7 What fall needs to be put in place for a 4.0 m length of guttering?

a 5.5 mm
b 6.6 mm
c 7.7 mm
d 8.8 mm

8 How long does a soundness test last for on a soil stack?

a 4 mins
b 3 mins
c 2 mins
d 1 min

9 What makes a shower trap different to the other styles of traps?

a It is smaller
b It is shallower
c It has a removable grate
d It is self-cleaning

10 What is the required trap seal depth for a basin discharging into a soil stack?
 a 38 mm c 62 mm
 b 50 mm d 75 mm

11 A 4.0 m length of plastic gutter is subjected to a 20°C temperature rise in the summer period. By how much will it expand if its coefficient of linear expansion is 0.06 mm/m/°C?
 a 2.4 mm c 4.8 mm
 b 3.7 mm d 5.3 mm

12 When carrying out a soundness test at a customer's property, which British Standard outlines the soundness test and performance test for a soil stack?
 a BS EN 806 c BS 1212
 b BS EN 12056 d BS 8000

13 Which one of these factors is not taken into consideration when designing a rainwater system?
 a Intensity of rain
 b Number of outlets
 c Roof material
 d Customer preference

14 Which item is installed at the base of a down pipe?
 a 45° angle connector c Swan neck
 b Shoe d Branch

15 What type of soil stack installation is this?

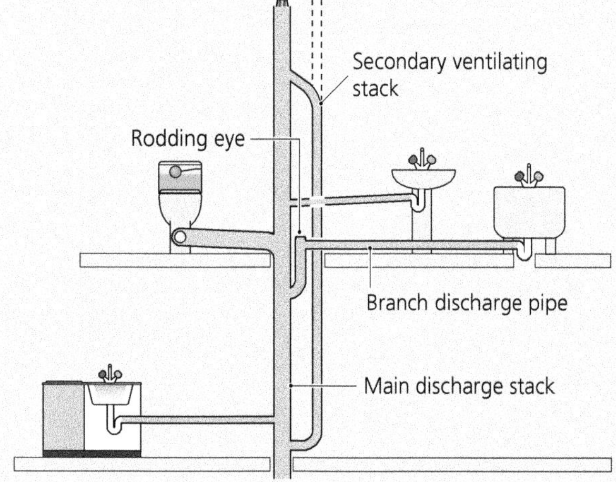

 a Primary ventilated
 b Additionally ventilated
 c Secondary ventilated
 d Ventilated discharge branch

16 What discharge pipe size would be connected to a single basin?
 a 32 mm
 b 40 mm
 c 50 mm
 d 110 mm

17 What style of WC pan is most commonly installed in domestic properties?
 a Single trap siphonic
 b Double trap siphonic
 c Wash down
 d Single close trap

18 Which of the following is not associated with a condensate drain from a condensing boiler?
 a Fall must be at least 2.5°
 b Any pipework outside must be as short as possible
 c It must not be made of metal
 d Minimum size is 40 mm

19 What discharge pipe size would be connected to a single bath?
 a 32 mm
 b 40 mm
 c 50 mm
 d 110 mm

20 Which one of the following statements does not describe a separate underground drainage system?
 a Cheap to install as there is less pipework
 b The rainwater and foul water enter different sewers
 c Cheaper in terms of water treatment
 d Care is required when connections are made

7 Health and safety and industry practices (Unit 211)

You will not be expected to have an in-depth knowledge of each of the practices in this section, but you do need to know what the particular legislations and documentation cover, how they are enforced and what your responsibilities are.

LO1 Understand health and safety legislation in the plumbing and heating industry

Topic 1.1 Types of health and safety guidance material

REVISED

Health and safety publications divide into two distinct groups: statutory and non-statutory.

Examples of statutory (mandatory) publications are:

- **Acts of Parliament** – these create or change the law. Example: Health and Safety at Work Act
- **Regulations** – these are rules and procedures set by government agencies.

Examples of non-statutory (**non-mandatory**) publications are:

- British Standards – are there for guidance
- manufacturers' instructions – are there for guidance.

The Health and Safety Executive (HSE) produces **HSE Guidance Notes** to help people understand what is required under the law.

There are also **Approved Codes of Practice** (ACoP) which help people understand what is required under a regulation.

Following the guidance notes, ACoP is considered good practice and makes sure you comply with the law.

All of the following 'Acts' and 'Regulations' are mandatory.

The Health and Safety at Work Act 1974 (HASAWA):

- The main piece of health and safety legislation.
- About people and activities, not the premises and processes.
- Everyone's responsibility.
- A health and safety policy must be produced if a company employs five or more employees.
- The employer must:
  - ensure the health, safety and welfare of the employees (as far as possible)
  - carry out risk assessments for work activities
  - implement control measures
  - inform employees of any risks and control measures
  - make a record of the risk assessments
  - supply correct PPE.

Mandatory This is the law (legal and legislation) which is enforceable and MUST be followed to avoid prosecution.

Comply Understand and put into action what is stated – observe, fulfil, confirm.

Exam tip

The titles of the Acts and Regulations give away what they cover.

- The employee must:
  - take reasonable care of the health and safety of themselves and others
  - comply with the health and safety policy
  - not recklessly interfere with anything that might affect health and safety
  - highlight any situation that may present serious or imminent danger.

Construction (Design Management) Regulations CDM:

- Improve planning and management of projects.
- Improve health and safety on site.

The Electricity at Work Regulations:

- Lay down requirements for safe working with electricity.
- The installer has responsibility to ensure that fixed electrical equipment is installed, tested and maintained correctly and portable equipment is regularly tested.
- They must be competent to undertake the work.
- Includes:
  - cable size, connections, fuse/MCB size, insulation (see Chapter 2 (Unit 213), LO5)
  - earthing, which keeps the user safe from electric shocks
  - PAT (Portable Appliance Testing) every three months.

Responsibility Having a duty towards something that you are accountable for.

Competent A person has been trained, has passed a test and has a certificate to prove they can carry out a task to a recognised standard.

Check your understanding

1 Who must complete the risk assessments as part of the control measures?

2 How often should you complete a PAT on a circular saw?

Control of Substances Hazardous to Health (COSHH) Regulations:

- Require employers to control exposure to hazardous substances to prevent ill-health.
- Include the use of solids, liquids, dust, fumes, vapours and gases.
- Training, instruction and information.
- Risk assessment of exposure.
- Control measures put in place.
- COSHH data folder kept.
- COSHH information is found on the outer box, inner container and an enclosed leaflet.
- COSHH symbols (global harmonised system of symbols) with a signal word. For example, TOXIC or FLAMMABLE (LO2 shows the symbols).

Work at Height Regulations:

- Apply to any work above ground level where there is a risk of a fall liable to cause personal injury.
- All work should be planned and organised.
- Weather conditions must be considered.
- Operatives must be trained and competent.
- Correct equipment chosen and inspected.
- Options other than working a height must be considered (such as bringing a component down to ground level to work on).

Personal Protective Equipment (PPE) at Work Regulations:

- PPE is defined as all equipment which is intended to be worn or held by a person which protects the person against one or more hazards.
- PPE is to be supplied and used wherever there is a risk to health and safety (for example, safety helmet, gloves, eye protection, high-visibility clothing, footwear).
- PPE should be accurately assessed before use to ensure suitability.
- PPE should be maintained and stored properly.
- PPE should be provided with instructions on how to use it safely.
- PPE should be used correctly.
- PPE must be supplied by the employer on a free of charge basis.

Manual Handling Operations Regulations:

- There is no safe weight to lift although there is guidance given.
- Kinetic lifting technique – knees bend with your back straight.
- Complete risk assessment prior to any lifting.
- Avoid lifting where possible.
- Take steps to reduce or remove the risk of injury.
- Use mechanical lifting aids when possible.
- When manually handling items, consider TILE: T – Task I – Individual L – Load E – environment.
- No matter who asks you, never lift heavy objects alone. Use a mechanical lifting aid or wait for help.

Check your understanding

3 What is meant by the kinetic lifting technique?

Provision and Use of Work Equipment Regulations (PUWER):

- Any risk to an operative's health and safety from work equipment should be prevented or controlled.
- PUWER:
  - Is the power tool suitable for its intended use?
  - Is the tool safe to use and maintained in safe condition?
  - Is the operative competent to use the tool?
  - Are suitable safety measures in place to use the tool?
  - Do any locks, guards and triggers work correctly?

Exam tip

PUWER – think of POWER (for example, power tools and their dangers when used).

Control of Asbestos at Work Regulations:

- White, blue and brown asbestos can only be removed by licenced contractors.
- White – chrysotile; Blue – amosite; Brown – crocidolite.
- Covered in more detail in LO2.3.
- Asbestos can only be disposed of by licenced contractors – double-bagged and labelled.
- If asbestos is suspected on site, work must be stopped and the material reported to your supervisor.
- Breathing in asbestos fibres can lead to chronic illness (mesothelioma or asbestosis).

Check your understanding

4 How must you dispose of asbestos safely?

Health, Safety and Welfare Regulations:

- These concentrate on the welfare of personnel.
- Include:
  - site access and walkways around the working area
  - mess huts, changing facilities, drying rooms
  - toilet facilities.

Exam tip

Health, Safety and Welfare Regulations – remember these by concentrating on WELFARE.

Health and Safety (First Aid) Regulations:

- These require employers to provide adequate and appropriate first aid equipment, facilities and personnel.
  - Competent first aider (trained, tested and has a certificate).
  - First aid room for larger sites.
  - First aid boxes (no medicines (tablets or creams) allowed in these boxes).

Confined Spaces Regulations:

- Avoid working in confined spaces if possible.
- Follow the safe system of work if confined space working cannot be avoided (LO7.2 outlines more detail).
- Adequate emergency procedures need to be in place before work starts.

Now test yourself TESTED

1 You are working on site and you come across something that looks like asbestos. What should you do?

Check your understanding

5 What would you not find in a first aid box?

Topic 1.2 Purpose of enforcing authorities and control measures

REVISED

Health and Safety Executive (HSE):

- They enforce the Health and Safety at Work Act – the principle legislation.
- Inspect sites, give guidance and advice, take photos, talk and discuss situations.
- If they find a breach of the HASAWA they can:
  - take informal action – give a word of advice
  - issue **improvement notices** – you get a period of time to put something right and the HSE will return to make sure all the required measures have been implemented
  - issue prohibition notices – work must stop until the problem is put right and the HSE will return to confirm
  - carry out powers of **prosecution** – if there is a failure to comply with legislation.

Prohibition Means that you must not do something.

The Local Authority:

- They employ Building Control Officers (BCO) who make sure the Building Regulations are being observed in the planning and building stage.
- BCOs have the power to reject plans.
- BCOs will visit sites at various times to ensure that Building Regulations are being observed.
- If Building Regulations are not being observed, the BCOs will refuse planning permission or not sign the work off.

Topic 1.3 Roles and responsibilities in relation to health and safety

REVISED

The Health and Safety at Work Act outlines that health and safety is everyone's responsibility:

- employers
- employees
- self-employed
- general public.

Table 7.1 Health and safety roles and responsibilities

Who	Health and safety roles and responsibilities

Employers	+ Health and safety of:
	  + their employees
	  + the general public
	+ Carry out risk assessments
	+ Produce Health and safety policy (if they employ five or more)
	+ Offer appropriate training

Who	Health and safety roles and responsibilities
Employees	+ Health and safety of: + themselves + fellow workers + the general public + Follow the risk assessments + Follow the Health and safety policy + Must not recklessly interfere with anything that may affect the health and safety of someone
Self-employed	+ Health and Safety of: + fellow workers + the general public + Create and follow the risk assessments + Must not recklessly interfere with anything that may affect the health and safety of someone

Designers, main and sub-contractors and clients also need to take the health and safety of everybody into consideration. They need to create risk assessments and work safely.

Now test yourself TESTED

2 You are working on site and there is an accident. The risk assessment needs to be updated. As an employee of your company, is it your responsibility to update the risk assessment?

Exam tip

Struggling to recall the regulation title and contents? Don't forget – the title gives the content away!

LO2 Understand hazardous situations within the plumbing and heating industry

Topic 2.1 Preventing potential site hazards

REVISED

Types of site hazards

There are many potential hazards on site, some more obvious than others. This is where common sense, good housekeeping and being aware of your working environment are important.

Housekeeping:

+ Water and oil make surfaces slippery.
+ Warning signs should be displayed.
+ Oil can be an irritant and sometimes carcinogenic.
+ Spillages can cause contamination.

Carcinogenic Something that can cause cancer.

Power tools:

+ PAT label in date.
+ Guard, lock and trigger working correctly.
+ Visual check.

Electrical:

+ Use a battery-operated tool if possible, as there is no chance of an electric shock and no trailing leads.
+ 110 V safe site voltage with yellow leads and plugs.
+ Use a residual circuit device (RCD) on mains-powered tools to keep you safe from electric shocks.
+ Don't overload cables as they could heat up and catch fire.

COSHH:

- Carry work out in a well-ventilated area or outside.
- Use extraction if possible where dust and fumes are present.
- Wear correct PPE.
- Never mix chemicals.
- Read the COSHH data sheets.
- Use chemicals in well-ventilated area or open windows.
- Be aware of asbestos.
- Chemicals and oil can cause skin irritation and contamination.

Fire:

- Waste material should be stored and disposed of correctly.
- Store materials safely.
- Fire extinguishers – water (red), foam (cream), CO_2 (black) and dry powder (blue).
- Don't overload cables as they could heat up and catch fire.
- Soldering – be aware of the flame, protect the customer's property. Stop soldering an hour before you leave the property.

Heights:

- Training is undertaken.
- Risk assessments are made and followed.
- Consider other options than just working at a height.
- Correct access equipment used along with harnesses.

Check your understanding

6 What does COSHH stand for?

The methods used to reduce the risk of injury from hazards

- Competent person's scheme – to make sure you are trained and tested.
- CSCS card system – to prove your level of responsibility on site.
- Permits to work – give controlled permission to carry out a task that carries a hazard (for example, a flame).
- **Risk assessment** carried out and followed:
  - identify the hazards and risks
  - control measures put in place
  - accident – unplanned, undesired event that may result in harm or loss
  - hazard – anything that has the potential to cause harm or loss
  - risk – the likelihood a hazard will actually cause harm.
- Use of PPE.
- **Method statement** or **safe work method** is part of the workplace safety plan.
  - Specific instructions on how to safely perform a work-related task.
  - Instructs employees on the approved way of working on site.
  - This may include a 'hot work permit'.
- Safety signs in place.
  - Prohibition: red circle, white background, black symbol with a red line across – you must not do something.
  - Mandatory: white circle, blue background with a white symbol – you must do something
  - Warning: black triangle, yellow background with a black symbol – warning you of a specific hazard.
  - Information: white square or rectangle, green background with a white symbol – used for emergency information like fire exits and first aid.
- COSHH symbols: red diamond, white background with a black symbol
  - Used to identify a potential hazard.
  - Part of a 'Global Harmonised System' (GHS) so they are recognised all around the world.

Exam tip

Struggling to recall signs and symbols? Look at the image on the sign that will help you identify it.

Topic 2.2 Types and characteristics of hazardous substances

REVISED

Lead:

- This is heavy and toxic and quite often installed at a height.
- Lead can be absorbed through the skin by touch.
- Lead can be ingested through the mouth.
- Lead can be inhaled through breathing.
- Wear barrier cream.
- Carry lead work out in a well-ventilated area or outside.
- Wash your hands carefully after use.

Toxic Means poisonous – it is also a COSHH symbol.

Solvents:

- Used to connect waste pipes together.
- They are highly flammable.
- The fumes can cause unconsciousness.
- They are bad for the environment.

Lubricants:

- Used to aid movement in tools and equipment.
- They are oil-based and so are a slip hazard if spilt.
- They can cause dermatitis – skin irritation and inflammation.
  - Use barrier cream.

Fluxes:

- Used in the soldering process to aid capillary action.
- There are two types of flux: active and inactive.
- Be aware that active flux is acidic and can burn the skin.
- Become sticky when heated and can cause burns.
- Create fumes when heated that should not be breathed in.

Jointing compounds:

- Used to help create watertight joints.
- Avoid skin contact.
- Read the manufacturer's instructions.
- Do not use oil-based compounds on potable water systems (contamination) or plastic components (degrading).

Sealants:

- Used in bathrooms, kitchens and toilets to form a watertight seal around an appliance to the wall or floor.
- These are a hazard to the environment.
- They create fumes that should not be breathed in.

Gases:

- Gases used on site can vary.
- Liquid Petroleum Gas (LPG): Propane (red bottle), Butane (blue bottle), MAPP gas (yellow bottle) and Acetylene (maroon bottle).
- They are heavier than air – danger if there is a leak.
- They are highly flammable.
- They produce hot flames.

Petroleum and diesel:

- Used to run site equipment.
  - Petrol-powered tools.
  - Diesel-powered generators.
- Both are flammable and hazardous to the environment.

Cleaning agents:

- Often water-based, but can be alkaline or acidic, some are solvent based.
- Read the COSHH data before using.
- Never mix chemicals as they can react and give off noxious gases.

Noxious Something (usually a gas) that is poisonous or very harmful.

COSHH symbols:

- TOXIC – poisonous.
- HARMFUL – warning that it has a hazardous effect.
- CORROSIVE – it will burn and corrode things away.
- IRRITANT – warning that it can cause skin inflammation.
- OXIDISING – it will burn and oxidise things away.
- FLAMMABLE – it can easily catch fire.

Now test yourself TESTED

3 You receive a new acid-based cleaning chemical. What COSHH symbols would you expect to see on the container?

Topic 2.3 Types and effects of asbestos exposure and how it should be prevented

REVISED

- There are three types of asbestos:
  - white – chrysotile
  - brown – amosite
  - blue – crocidolite.
- Asbestos is a fibrous material found in properties built between 1940–1980, but care should be taken in properties built up until 2000.
- If you suspect that asbestos is present in the building, stop work immediately and report it to your supervisor. The asbestos will need to be removed to protect the workforce and public.
- Before work begins, there will be a building inspection where material samples should be taken and logged in the register of materials.
- Asbestos could be found in flues, guttering, pipe insulation, roofing, boiler gaskets, cisterns, artex, heat proof material, soil and rainwater pipes.
- Any asbestos needs to be removed and disposed of by licenced contractors, double-bagged and labelled. It has to be removed by a licenced contractor because asbestos is a hazardous waste.
- Asbestos can cause **mesothelioma** or asbestosis which are both chronic (long-term) illnesses.

Mesothelioma A form of lung cancer (chronic illness).

Topic 2.4 Types of waste management and disposal

REVISED

- It is important to try to minimise waste from site
- Recycling is an important form of waste management.
  - Material can be reformed to be reused – cardboard, plastics and metals.
  - Helps to save the environment.
  - Get money back for recycled material like copper, brass, steel, aluminium.
  - Recycle skips on site.

- Landfill is used for any waste that cannot be recycled.
  - Not good for the environment.
- Electrical equipment has to be disposed of under the Waste Electrical and Electronic Equipment (WEEE) Regulations. This minimises waste and maximises re-use where possible.
- Any waste has to be disposed of correctly and business waste should be collected by a company that has a waste carrier's licence and will dispose of the waste correctly.
- Asbestos waste must be removed and taken by an approved/licenced contractor as with any other hazardous waste.
- We must be aware of various forms of contamination that we may cause:
  - land and air contamination from chemicals and fumes
  - noise and light contamination either temporarily when installing (for example, power tools, radios, security lighting) or permanent from an installation (for example, fan on an air source heat pump or AC unit).

Exam tip

Don't forget that contamination and pollution can take various forms (for example, chemicals, fumes, light, noise).

LO3 Use personal protection and respond to accidents

Topic 3.1 Use PPE for plumbing and heating work

REVISED 

- PPE is designed to protect against workplace hazards.
- PPE is required to be supplied by your employer free of charge.
- It is your responsibility to wear the PPE correctly, maintain it in good condition and report any problems.

This table is important as questions will be asked about identification and when items are used.

Table 7.2 PPE

Part of body	Hazards	Options
Eyes	Chemical or metal splash, dust, projectiles, gas and vapour, radiation	Safety spectacles, goggles, face screens, face shields, visors
Head and neck	Impact from falling or flying objects, risk of head bumping, hair getting tangled in machinery, chemical drips or splash, climate or temperature	Industrial safety helmets, bump caps, hairnets and firefighters' helmets
Ears	Noise – a combination of sound level and duration of exposure; very high-level sounds are a hazard even with short duration	Earplugs, earmuffs, semi-insert/canal caps
Hands and arms	Abrasion, temperature extremes, cuts and punctures, impact, chemicals, electric shock, radiation, vibration, biological agents and prolonged immersion in water	Gloves, gloves with a cuff, gauntlets and sleeving that covers part or all of the arm
Feet and legs	Wet, hot and cold conditions, electrostatic build-up, slipping, cuts and punctures, falling objects, heavy loads, metal and chemical splash, vehicles	Safety boots and shoes with protective toecaps and penetration-resistant mid-sole, wellington boots and specific footwear (such as foundry boots and chainsaw boots)
Lungs	Oxygen-deficient atmospheres, dusts, gases and vapours	Respiratory protective equipment (RPE)
Whole body	Heat, chemical or metal splash, spray from pressure leaks or spray guns, contaminated dust, impact or penetration, excessive wear or entanglement of own clothing, falling from heights	Conventional or disposable overalls, boiler suits, aprons, chemical suits, a full body harness may be used when working at height

Typical mistake

Relating the item of PPE, what hazard it protects the user from and the work activity undertaken. In the exam, you might be asked what would be the **PRIMARY** piece of PPE that is needed. Think about the major hazard and the required PPE to protect against that hazard (for example, the main PPE required when drilling a hole would be eye protection, even though you may be required to wear other forms of protection).

Topic 3.2 Perform manual handling

REVISED

This is a workshop activity in which you will need to show that you know how to perform manual handling, including the use of the following:

- manual lifting technique
- mechanical aids
- load assessment.

Topic 3.3 First aid in plumbing and heating industry

REVISED

- An accident is an unforeseen, unplanned and uncontrolled event.
- Accidents can happen anywhere. Remember, we work in small, unoccupied properties; occupied properties; construction sites; and we travel on the road.
- First aid boxes must be located on site and in any vehicle going on site.
- No medicines (creams or tablets) must be in a first aid box.
- A first aider will be on site – competent person who has been trained, tested and has a certificate.
- The first aider will take charge of a situation, check first aid boxes and be on call while on site.

Accident An unforeseen, unplanned and uncontrolled event that can damage something or injure someone.

Topic 3.4 Dealing with accidents on a construction site

REVISED

If an accident or emergency occurs, time is important.

Raise the alarm:

- Dial 999.
- Speak clearly to the operator.
- Describe the nature of the incident and the location.
- Get someone to wait at the site entrance to direct the services.
- Stay with the injured person (unless there is a danger to your life).
- Allow the first aider to take charge, who will hand over to the emergency services.

Evacuation:

- If there is a fire or emergency, the site will need to be cleared.
- Sound the alarm (for example, hooter, siren, bell, whistle, shout).
- Call 999.
- Speak clearly and describe the situation and location.
- Walk carefully off-site, following the emergency exit route.
- Meet at the designated assembly point.
- Trained staff may attempt to deal with the situation if safe to do so.

Reporting accidents:

- Record every injury, even minor injuries, in the accident book.
  - Name and address.
  - Date of accident.
  - Details of accident and injury.
  - Cause.
  - Name and address of witnesses.

- The Reporting of Injuries, Diseases and Dangerous Occurrences Regulations (RIDDOR)
  - If there is a serious injury, near miss or fatality, it must be reported to the HSE.
  - If a worker is off work for more than five days, it must be reported within 15 days to the HSE.
  - If there is a fatality, it must be reported immediately.

Injuries:

- Minor cut – clean, stop bleeding, bandage.
- Minor burn – cool area (10 mins under water), clean, bandage.
- Objects in the eye – eye wash, medical treatment if more serious.
- Fume exposure – remove from fumes into well-ventilated area, recovery position, CPR method.
- Bone fractures:
  - Simple – broken bone no wound
  - Compound – broken bone with wound.
  - Check breathing.
  - Check for deformity.
  - Support injury.
  - Reassure patient.
  - Medical treatment.
- Electric shock – take extreme care, isolate electric source, CPR method, recovery position.
- Unconsciousness – loosen clothing, lie patient on their back, check breathing, CPR method, recovery position.
- Concussion – loosen clothing, lie in recovery position.
- Recovery position – patient on their side with airways open.

RIDDOR Reporting of Injuries, Diseases and Dangerous Occurrences Regulations.

CPR (cardiopulmonary resuscitation) A method to keep someone alive in a medical emergency.

Now test yourself TESTED

4 You are working on site and a fire starts near to a small wood store. What is the first thing you should do?

Check your understanding

7 If you burn your hand on a freshly soldered joint, how long should you cool it for?

Typical mistakes

Struggling to recall the correct first aid treatment for minor injuries. Remember: cuts (clean and stop bleeding), burns (cool under water for 10 mins) and exposure (fresh air).

LO4 Understand procedures for electrical safety

Topic 4.1 Types of electrical supplies used on site

REVISED 

Electric shocks are a major hazard and great care should always be taken when working with electricity.

Battery-powered:

- Battery-powered tools are preferred.
- No trailing leads.
- Flexible.
- No electric shock from 18 V.

110 volt:

- Site safe voltage.
- Yellow lead and plug.
- May require a step-down transformer.
- Reduced chance of injury from electric shock.

230 volt:

- Domestic voltage.
- Risk of electric shock.
- Blue lead and plug on site.

Topic 4.2 Types of electrical hazards and safety

REVISED

- Always take great care when working with electrical power tools.
- Do visual inspections.
- Check for signs of damage and wear (such as worn cables).
- Check there are no trailing cables and/or extension leads.
- Check there are no cables near to pipework.
- Check for any hidden/buried cables.
- Check for wrong overcurrent protection devices (for example, fuse or MCB too small/big).
- Use temporary continuity bonds:
  - when cutting into pipework
  - to prevent electrical shocks.

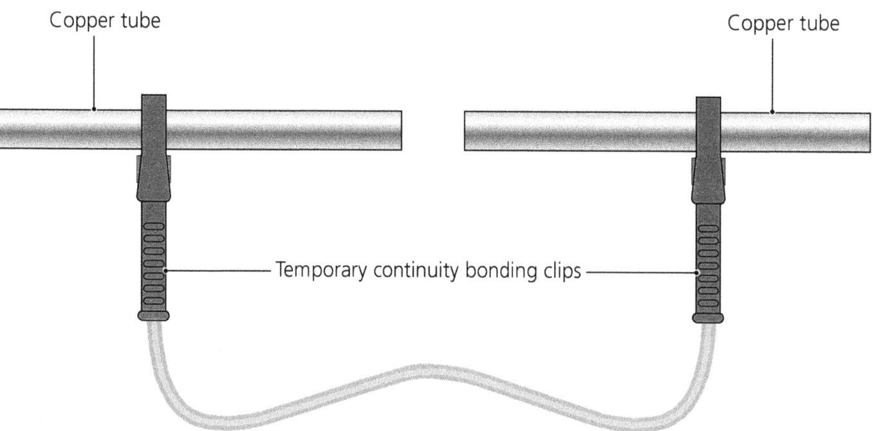

Temporary continuity bonds

- Check PAT certificate is in date.
  - Every three months and, if hired, every time the tool is returned.
- Use RCD protection on 230 V power tools.
  - Earth to live fault.
  - Protects from electric shock.

Temporary continuity bonds Used when cutting into pipework to protect against electric shock (they look like jump leads).

RCD (residual current device) A device that quickly breaks an electric circuit.

Check your understanding

8 What sort of power tool would you choose to use in order to avoid as many hazards as possible in a customer's property?

Topic 4.3 Safe isolation procedure

REVISED

The safe isolation procedure **must** be carried out before working on any electrical appliance.

1. Identify the circuit that needs to be isolated.
2. Let customer know isolation is going to be carried out.
3. Use voltage test indicator to test circuit is live.
   - Live to earth 230 V.
   - Live to neutral 230 V.
   - Earth to neutral 0 V.
4. Isolate circuit at fuse/MCB – lock and label.

Voltage test indicator Indicates the presence of voltage (dead test – no voltage; live test – voltage and lights come on).

5 Use voltage test indicator to re-test circuit is dead.
   - Live to earth 0 V.
   - Live to neutral 0 V.
   - Earth to neutral 0 V.
6 Re-test voltage indictor on known source (proving unit) – lights come on.
7 Start work.

- Central heating system isolated at switch spur.
- Electric shower isolated at the consumer unit MCB.
- Immersion heater isolated at the consumer unit MCB.

Proving unit Shows the voltage test indicator is working.

Exam tip

It is important that you are able to put the sequence of safe isolation in the correct order. Safe isolation comes up in Chapter 2, Electrical and scientific principles (Unit 213), Chapter 3, Cold water (Unit 214), Chapter 4, Hot water (Unit 215) and Chapter 5, Central heating (Unit 216) – so it is important to learn it!

Typical mistake

Struggling to recall the correct safe isolation procedure and relate it to a working scenario (for example, before replacing an electric shower, what should you do?). If the appliance requires an electrical supply to work, it must be safely isolated prior to work beginning to avoid an electric shock.

LO5 Work with heat producing equipment

Topic 5.1 Gases used in equipment

REVISED

When working with any liquid petroleum gas (LPG), you should have a CO_2 fire extinguisher present.

Propane:

- Red cylinder.
- Highly flammable (remember COSHH symbol).
- Heavier than air (store outside, well-ventilated area, away from drains).
- Turns from liquid to gas at –42°C.
- Used for soldering.

Butane:

- Blue cylinder.
- Highly flammable (remember COSHH symbol).
- Heavier than air (store outside, well-ventilated area, away from drains).
- Flame temperature too high for soldering, used on BBQ.

Oxy-acetylene:

- Oxygen – black cylinder.
- Acetylene – maroon cylinder (highly flammable).
- Lighter than air (store outside, well-ventilated area).
- Very hot flame temperature.
- Used in commercial pipe welding.

Nitrogen:

- Grey cylinder.
- Non-flammable.
- Used in welding, testing and as a medical gas.

Transportation:

- When transporting LPG, you will need the Hazchem (COSHH) symbol on show.
- If transporting large quantities of LPG, you will need the 'International Carriage of Dangerous Goods by Road' certificate.

Now test yourself TESTED

5 You are sent to the LPG store on site to collect a replacement cylinder so the team can continue soldering copper pipe together. What colour cylinder would you look for?

Check your understanding

9 What colour cylinder is propane stored in?

Topic 5.2 Fire safety principles

REVISED

- Fire is very destructive – it is a chemical process.
- As plumbers we use fire every day, so care is needed.

Fire triangle:

- You need all three elements for combustion.
- Take one away and the fire goes out (that's how a fire extinguisher works).
- Equations:
  - Oxygen + Fuel + Heat = Fire
  - Oxygen + Propane + Spark = Flame (for soldering)

Fire triangle

Classification of fires:

- Class A – solid material, wood, paper, textiles.
- Class B – flammable liquids.
- Class C – flammable gases.
- Class D – metals.
- Class E – electrical.
- Class F – cooking fat/oil.

Exam tip

Remember A – B – C = Solid – Liquid – Gas (to help you remember the classes of fires).

Typical mistake

Struggling to recall the class and type of extinguisher used. Learn the information in Table 7.3 to help you.

Types of fire extinguishers, identification colours and uses:

Table 7.3 Types and uses of fire extinguishers

Class A fire	Class B fire	Class C fire	Class D fire	Class E fire	Class F fire
WATER					Special chemical extinguisher
FOAM	FOAM				
DRY POWDER	DRY POWDER	DRY POWDER	DRY POWDER	DRY POWDER	
	CO_2	CO_2	CO_2	CO_2	

- Dry powder extinguishers make a mess.
- NEVER use a water/foam extinguisher on an electrical fire.

Evacuation procedures:

- If there is a fire, the site will need to be cleared.
- Sound the alarm (for example, hooter, siren, bell, whistle, shout).
- Call 999.
- Speak clearly and describe the situation and location.
- Walk carefully off-site, following the emergency exit route.
- Meet at the designated assembly point.
- Trained staff may attempt to deal with the situation if safe to do so.

Exam tip

If asked what fire extinguisher a plumber needs to keep nearby when soldering, it is a CO_2 extinguisher.

Now test yourself TESTED

6 A small fire breaks out in the wood store on site. There is a choice of fire extinguishers nearby. What colour band extinguisher and content would you look for?

Topic 5.3 Assemble LPG equipment

REVISED

This is a workshop activity in which you will need to show that you know how to assemble LPG equipment, including flammable gas cylinders.

- These have a left-hand thread – for safety.
- Inspect all equipment for damage.
- Connect parts together.
- Connect flashback arrestor – prevents a flame from entering a cylinder.
- Full procedure is in the C&G Plumbing: Book 1 textbook.
- Test for leaks using leak detection fluid (NEVER use a naked flame).
- Light acetylene first, then add the oxygen.
- Turn the acetylene off first, then turn the oxygen off.

Flashback arrestor Prevents a flame from entering a cylinder (stops it flashing back down the hose to the cylinder).

LO6 Use access equipment on a construction site

Topic 6.1 Types of access equipment

REVISED

Steps:

- Low-level access to ceiling height.
- Inspect before use – feet, steps, platform and so on.
- Make sure locks are engaged.
- Step-to-face work activity.
- Electrician's stepladders are made of fibreglass – coloured yellow.

Ladders:

- Extension and pole ladder (one piece).
- Class 1 (site work), Class 2 (professional use), Class 3 (DIY).
- Pole ladder – scaffold access.
- Extension ladder – high-level inspection (gutter).
- Do not lean – 3-point contact – secure at top – 30 minutes of work only.
- One person at a time.
- 75° angle or 4:1 ratio – be prepared to calculate this!
  - 4 metres up: 1 metre out.
  - 5 metres up: 1.25 metres out.
  - 6 metres up: 1.5 metres out.
- Stand-off used to protect gutter.
- Extend five rungs or 1.0 m above working height.

Mobile scaffold tower:

- High-level work (gutter or soil stack).
- Follow manufacturer's instructions.
- Lock wheels, and point wheels in at a 45° angle.
- Out-riggers for stability.
- Do not move with tools or people on.
- Working platform with toe board and guard rails.

Platforms:

- Step up.
- Trestles.
- Low-level access.

Harnesses:

- Used for high-level, exposed work.
- Training required.
- Never work alone.

Stand-off An attachment for a ladder to enable the ladder to 'stand off' the building, so that items like guttering are not damaged.

Exam tip

Be prepared to calculate both heights and distances from the base of a wall/ platform. For example, you might be asked: 'If a working platform is 6.0 m high, how far from the base of the platform should the base of the ladder be?' or 'If the base of a ladder is 1.5 m from the base of a working platform, how high is the platform?' These questions are asking about the same situation.

Roof ladders and crawling boards:

- Use to avoid damaging roof tiles.
- Wheels ride up tiles.
- Roof ladders hook over roof ridge.
- Access to roof line working (flues, lead flashing).
- Crawling boards used for access over fragile areas (unboarded lofts).

Fixed scaffolds and edge protection:

- Used when building a property or working on a larger area of a property.
- Edge protection is there to prevent the operative falling off the edge of an exposed area.

Mobile elevated platforms:

- Scissor lift.
- Cherry picker.

With all access equipment:

- Inspect.
- Training.
- Weather conditions.
- Be aware of overhead cables.
- Ground conditions – wet, slippery, level, firm.

Exam tip

If the word 'inspection' is used in a question, this means very short-term work is carried out. If the word 'replacement' or 'rectification' is used in a question, this means longer-term work is carried out. The access equipment could change accordingly.

Topic 6.2 Use access equipment

REVISED

This is a workshop activity in which you will need to show that you know how to use access equipment, including the use of the following, with all necessary pre-checks before using:

- steps
- ladders
- mobile scaffold towers
- platforms.

LO7 Understand how to work safely in excavations and confined spaces

Topic 7.1 Working practices in excavations

REVISED

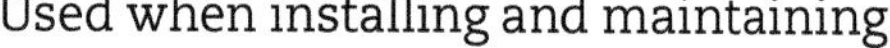

Used when installing and maintaining:

- cold water mains to a property
- below ground drainage
- sewage pipes
- rainwater harvesting equipment.

Main dangers:

- Collapse of the side of the excavation.
- Falling objects, like tools, soil or rocks.
- Flooding.

Need to be planned and managed:

- Excavations deeper than 1.2 m need supporting.
- Care with type of ground: sand is soft – clay is hard.
- Ladder access and ladder emergency exit.
- Warning signs.
- Barriers.
- Vehicle proximity and stop blocks (vibration could crumble sides).

Topic 7.2 Working practices in confined spaces

REVISED

Examples:
- duct work
- plant rooms
- tanks, cylinders, boilers and cisterns
- under suspended floors
- roof spaces
- rainwater harvesting systems
- wells.

Main dangers:
- Lack of ventilation (lack of oxygen).
- Poor lighting.
- Flooding.
- Obstruction of escape route.
- Collapse of the structure.

Need to be planned and managed:
- Identification of responsibility.
- Precautions.
- Risk assessments.
- Training and instruction.
- Ensure you never work alone.

Typical mistake

Not realising the question is relating to a confined space or excavation. For example, the question might just say you are working in a loft or installing a cold water main to a property.

Now test yourself

TESTED

7 As part of a new build project, your company has been asked to connect the cold water system of a house to the water meter at the property boundary. What would be a primary concern when installing the MDPE pipework?

Exam-style questions

1 Which one is not an immediate danger when replacing the cold water main to a customer's property during an excavation?
   a Collapse
   b Explosion
   c Flooding
   d Falling objects

2 Which one of the following is not a danger to the operative when working in a sewer pipe underground?
   a Lone working
   b Respirator
   c Flooding
   d Weil's disease

3 Which one of the following situations would not be classified as a confined space when working in that location?
   a Outside toilet
   b Loft area
   c Well
   d Rainwater harvesting cistern

4 If you extend your ladder to 5.0 m, how far from the base of the wall does it need to come out?
   a 1.0 m
   b 1.25 m
   c 1.5 m
   d 1.75 m

5 Identify the piece of access equipment that gives vertical, as well as mobile, access to a very high level.
   a Step ladder
   b Access cage
   c Scissor lift
   d Extension ladder

6 You are asked to erect an extension ladder against some existing scaffolding. Where is the preferred place to secure the ladder?
   a Base
   b Part-way up
   c Top
   d Anywhere

7 You have to get access to a cold water storage cistern in a customer's loft, but the loft is not boarded. Which access item would need to be used?
   a Extension ladder
   b Step up
   c Trestle
   d Crawling board

8 When putting together a mobile scaffold tower, what do the out-riggers do?
- a Allow a larger working platform
- b Increase stability
- c Enable heavy tools to be used
- d Allow more than one person to work on the platform

9 You have been asked to go around to a customer's property where a guttering joint, at first floor height, is leaking. What access equipment would you use to inspect the situation?
- a Stepladder
- b Extension ladder
- c Mobile scaffold tower
- d Roof ladder

10 You are working on a multi-trade site and need to use a stepladder, but your stepladder is in the van. You ask someone if you can borrow their stepladder for a minute. You are handed a stepladder that is different to your normal one. Which of the following would be wrong?
- a The stepladder is yellow
- b It is extendable
- c You borrowed it from an electrician
- d It is made from fibreglass

11 When working on site a sub-contractor asks you if it is okay to solder copper pipework using a blowtorch and red bottle. What does the red bottle contain?
- a Butane
- b Acetylene
- c Oxygen
- d Propane

12 Which item is missing from the fire triangle if fuel and oxygen are present?
- a Heat
- b Nitrogen
- c Paper
- d Air

13 Which type of fire extinguisher must not be used on electrical fires?
- a CO^2
- b Dry powder
- c Red
- d Black

14 You are about to use an oxy-acetylene blowtorch. As part of the induction. your supervisor asks you a couple of safety questions. Why is the flashback arrestor so important?
- a Prevents a flame entering the blowtorch
- b Prevents a flame entering the hoses
- c Prevents a flame entering the gauges
- d Prevents a flame entering the cylinders

15 You are working in a customer's property and you suspect your propane gas torch is leaking. Which of the following must you not do?
- a Take it outside
- b Put it in the van and get a new one
- c Test the connections with leak detection fluid
- d Tighten up any connections

16 What would be the preferred choice of power tools when drilling a hole in the wall at a customer's house?
- a SDS battery powered drill
- b SDS 230 V power drill
- c 110 V rotary drill
- d 400 V power drill

17 When working on site, a new apprentice starts and asks what colour the safe site voltage plug is. What is the correct answer?
- a Blue
- b Red
- c Black
- d Yellow

18 What is the most important thing to do before using power tools?
- a Put safety gloves on
- b Carry out a visual inspection of the tool
- c Clean the tool from when it was last used
- d Carefully place a dust sheet on the floor

19 When replacing an electric shower at a customer's property, you need to make sure the safe isolation procedure is followed. Which of the options contains part of the safe isolation procedure to carry out this work?
- a Isolate the mains isolator at the consumer unit and start work straight away
- b Lock and label MCB for the circuit
- c Check appliance lights are off
- d Start work straight away

20 Which is the correct course of action to treat a minor burn?
- a Take to hospital and get treated
- b Carefully cover with a bandage to protect the injury
- c Cool under running water for 10 minutes
- d Place the patient in the recovery position

21 An emergency happens on site and you need to call the emergency services. What detail will need to be conveyed?
- a Your home address
- b Your name and telephone number
- c A brief description of the emergency
- d Details of the first aid treatment

22 An accident occurs on site and you were present to initially help with the situation. What does not need to take place after the patient has been taken to hospital?
- a Make a witness statement
- b Complete the accident book
- c Complete a COSHH form
- d Inform the next of kin

23 When drilling a hole in the wall for a new boiler flue, you use an SDS drill with a diamond core bit. What PPE would you be required to wear?
- a Hi-visibility jacket, safety shoes, eye protection, knee protection
- b Hard hat, respirator, safety shoes, gloves
- c Ear protection, dust mask, eye protection, hard hat, gloves
- d Eye protection, hard hat, safety shoes, hi-visibility jacket

24 You are working on site and have been asked to display a circular sign with a red outline, the image of a person walking and a red line crossing through the image. When would be the correct time to display this sign?

- **a** In the hours of darkness
- **b** Only at times when the public have access
- **c** At all times
- **d** Only if there are groups of more than three people

25 What does this sign mean?

- **a** Flammable
- **b** Corrosive
- **c** Harmful
- **d** Oxidising

26 What regulations relate to the disposal of electrical equipment?

- **a** COSHH
- **b** WEEE
- **c** RIDDOR
- **d** HASAWA

27 The skin on your hand becomes red and inflamed and starts to itch. Which of the following operations could have caused dermatitis?

- **a** Tightening a compression fitting
- **b** Replacing a propane cylinder on a blow torch
- **c** Cutting a thread using cutting oil as a lubricant
- **d** Drilling a hole in a wall using an SDS hammer drill

28 Who must provide personal protective equipment for site work?

- **a** The site foreman
- **b** Main contractor
- **c** The client
- **d** The employer

29 You accidentally cut your finger when working on site. How should you treat a minor cut?

- **a** Wash for 10 minutes until clean
- **b** Sterilise with antiseptic cream
- **c** Clean and cover
- **d** Use CPR

30 You need training to become competent in the use of a circular saw. What regulation covers this requirement?

- **a** Health, Safety and Welfare Regulations
- **b** Personal Protective Equipment at Work Regulations
- **c** Provision and Use of Work Equipment Regulations
- **d** Control of Substances Hazardous to Health Regulations

Exam breakdown

Test specification

The new technical qualifications have been produced to meet the new level of rigour and robustness for vocational qualifications. One of these changes is the increased emphasis on the external assessment/exam.

The qualification includes an externally set and marked exam, which is taken at the same time by all candidates nationwide. City & Guilds produce the exam timetable. All your practical and theory training will be given to you in time to meet this date and time.

Self-motivated and independent study will be important to back up the input from your training provider.

The purpose of the assessment/exam is to prove that you have achieved sufficient knowledge and understanding from your study and that you can independently recall and draw on that knowledge and understanding.

Exam structure

Assessment type: 60 multiple-choice questions

Duration: 120 minutes

Access: Paper-based or online.

You can only re-sit this exam **once**. The first exam date will be in March and the second re-sit date will be in June of the same year. If you fail to pass both attempts, you will not achieve the qualification in that academic year.

The 60 questions will cover all the theory areas and are broken down as follows:

- 29 questions on basic knowledge (48%)
- 19 questions testing your understanding (32%)
- 12 questions applying your knowledge (20%)

Example questions

Some sample questions are given below. *Indicates the correct answer.

Note the different styles of questions!

Knowledge

What are ripples **most** commonly caused by when bending copper pipe in a bending machine?

a) Excessive pressure from the roller
b) Inadequate pressure from the roller*
c) Using new equipment
d) Using old equipment

Understanding

What size of current protection device is most suitable to protect a circuit that has a 9 kW electric shower installed on a 230 V supply?

a) 16 amp
b) 32 amp
c) 40 amp*
d) 50 amp

Application

A ground-floor flat in a five-storey block is having problems with foam and waste water appearing in the bath and WC, which are connected to the soil stack. The inspection chamber is checked, but there is no blockage. What is the most likely reason this is happening?

a) Incorrect type of trap being used on the appliances
b) Waste pipes enter the stack at the incorrect angle
c) Too many washing machines and dishwashers are connected to the stack
d) Connection at the base of the stack is less than 750 mm to the invert*

Exam specification

Unit	Title	Number of questions
211	Health and safety	8
212	Plumbing processes	6
213	Electrical and scientific principles	8
214	Cold water	6
215	Hot water	7
216	Central heating	6
217	Sanitation and drainage	7
Integration across all units		**48 questions**
Applied knowledge and understanding		**12 questions**
Total		**60 questions**

The integration questions are the stretch and challenge questions to differentiate your performance in knowledge and understanding.

Tackling questions

When you go in for the exam you really need a clear head to tackle the questions that will be asked. Make sure you have slept well leading up to the exam and have eaten a good meal on the morning of the exam.

When you start the exam, make sure you have the right attitude to get the best result you possibly can. Do not go into the exam wanting to get out as quickly as possible; use the time given to you effectively – 2 hours to answer 60 questions.

When you look at each question, it is so important to read the **whole** question and **understand** what is being asked. Then **think** about what the answer could be **before** looking at the choice.

This strategy may be helpful:

1 **Read** the whole question.
2 **Understand** what is being asked.
3 **Think** about the possible answer.
4 **Look** for the answer.

So, let's look at a couple of examples.

Here is short question that you could so easily not read fully or properly understand.

Examples 1 and 2

Q1 What type of access equipment would be used to access some guttering for inspection?

Q2 What type of access equipment would be used to access some guttering that needed replacing?

If the question was not read completely, you might read: *What type of access equipment would be used to access some guttering?* for both questions!

The key to each question is in the final part: *'for inspection'* or *'that needed replacing'*.

The final part would indicate two different types of access equipment:

+ 'for inspection' – short-term work, just looking, therefore, a ladder would be used.
+ 'that needed replacing' – longer-term work, installing components, therefore, a mobile scaffold tower would be used.

This means do not rush to answer before you have read the question **fully** and **thought** about the answer.

Example 3

Q3 Under the requirements of the Control of Substances Hazardous to Health Regulations:

a) hazardous materials should not be used in enclosed spaces such as those encountered in buildings
b) persons using materials hazardous to health must be provided with proper information, instruction and training
c) hazardous materials must not in any circumstances be used by anyone under the age of 18 years of age
d) the intention to use a hazardous material must be advised to the Health and Safety Executive.

Now, with this question, the answers are longer and need to be correctly read.

Using the strategy above, the first thing you need to do is:

1 **Read** the question fully.

Having read the question, you need to:

2 **Understand** what is being asked.

This question is about the fundamentals of COSHH – the Control of Substances Hazardous to Health.

Now:

3 **Think** about COSHH and mentally list what you know; for example:

+ Keeping safe
+ Training
+ Information
+ COSHH folder
+ PPE
+ Symbols

Once you have made that mental list:

4) **Look** for the answer containing those items, and that is option B.

Example 4

Q4 There is no supply of water to the cold water tap on the bath when it is opened. The cold water system is indirect. The MOST likely cause of the problem is fault on the:

a) service valve to the cold water storage cistern
b) cold feed supply pipework
c) cold distribution pipework
d) stop valve to the rising main

1 **Read** the question fully.
2 **Understand** what is being asked.

There is no cold water coming out of the bath tap, so where could the fault lie? This means you have to think about the system layout.

3 **Think** about the system. Picture it in your mind, or even draw it out on a spare piece of paper. Being an indirect cold water system, the cold water comes from the cold water storage cistern through the cold distribution and out to the tap.
4 **Look** for the answer. Which answer suggests the cold distribution? Option C.

Practise this strategy on some of the questions at the end of each unit, especially a unit that you may find more difficult, and try your hardest to get the best mark you can!

Good luck!

Glossary

Accident An unforeseen, unplanned and uncontrolled event that can damage something or injure someone. Page 152

Carcinogenic Something that can cause cancer. Page 147

Centralised Hot water delivered from a central point to the outlets (cylinder or combination boiler). Page 86

Cleat (or noggin) A piece of wood positioned to support the replaced floor. Page 19

Coefficient of linear expansion All materials expand by a small amount in length when heated. The amount they expand by is measured in millimetres. This number is known as the coefficient of linear expansion: it is a measure of how much a material expands by for every degree C it heats up, per metre of pipe or material used. The heat could be caused by many things including the sun, a flame, or hot water passing through.

You can calculate this by:

length of pipe × coefficient of expansion × temperature rise.

So, for 5.0 m of plastic guttering in summertime when the temperature rises 150C: 5000 mm × 0.00018 × 15 = 13.5 mm expansion. Page 33

Competent A person has been trained, has passed a test and has a certificate to prove they can carry out a task to a recognised standard. Page 144

Comply Understand and put into action what is stated – observe, fulfil, confirm. Page 143

Condense When molecules move together and form water droplets. Page 58

CPR (cardiopulmonary resuscitation) A method to keep someone alive in a medical emergency. Page 153

DZR The letters used on new brass fittings to identify de-zincification resistance. Page 34

Equilibrium Balanced. Page 42

Estimate An approximate price that could vary slightly. Page 17

Evaporate When water molecules move apart and turn to gas. Page 58

Flashback arrestor Prevents a flame from entering a cylinder (stops it flashing back down the hose to the cylinder). Page 157

Fulcrum The hinge point for a lever. Page 44

Grommet A rubber seal used on the cold water storage cistern. Page 66

Impeller An internal rotating paddle that powers the water in a pump. Page 101

Instantaneous Water heater on demand to the outlets (combination boiler or thermal store). Page 84

LCS (Low Carbon Steel) Used for commercial pipework. Page 10

Localised Hot water heated and delivered at the point of use to the outlet (single point water heater). Page 84

Magnetite A form of rusting or oxidation on the inside of the central heating system. Also known as 'black sludge'. Page 36

Mandatory This is the law (legal and legislation) which is enforceable and MUST be followed to avoid prosecution. Page 143

MCB Micro Circuit Breaker or Mini Circuit Breaker. Page 50

Mesothelioma A form of lung cancer (chronic illness). Page 150

Noxious Something (usually a gas) that is poisonous or very harmful. Page 149

Open vented A system open to atmospheric pressure by the use of an open vent pipe. Page 82

Precipitation Water that falls from clouds to the ground, such as rain, snow, sleet. Page 58

Prohibition Means that you must not do something. Page 146

Proprietary fitting A fitting 'made for the purpose of' something. Page 25

Proving unit Shows the voltage test indicator is working. Page 155

PTFE tape Polytetrafluoroethylene tape (or plumber's tape for everything). Page 29

Quotation A fixed price that cannot vary. Page 17

RCD (residual current device) A device that quickly breaks an electric circuit. Page 155

Responsibility Having a duty towards something that you are accountable for. Page 144

RIDDOR Reporting of Injuries, Diseases and Dangerous Occurrences Regulations. Page 153

Room sealed Where the boiler draws air direct from outside of the building through the same flue used to discharge the combustion gases. Page 106

S trap A style of trap used under a sanitary appliance and derives its name from its shape. Page 37

Stagnation When water is allowed to stand still and becomes stale and foul. Page 66

Stand-off An attachment for a ladder to enable the ladder to 'stand off' the building, so that items like guttering are not damaged. Page 157

Stored Water heated and kept prior to demand (hot water cylinder). Page 84

Syncron motor An electric motor located inside a two-port valve that moves the paddle and engages the micro-switch. These can be replaced easily. Page 100

Temporary continuity bonds Used when cutting into pipework to protect against electric shock (they look like jump leads). Page 154

Toxic Means poisonous – it is also a COSHH symbol. Page 149

Tundish Part of the discharge pipework joining D1 and D2 together, offering a visual sight of any discharge. Page 88

Unvented A sealed pressurised system with safety controls. Page 88

Voltage test indicator Indicates the presence of voltage (dead test – no voltage; live test – voltage and lights come on). Page 154

Picture credits

We would like to thank City & Guilds for permission to reuse artworks from their Plumbing/Electrical Installations textbooks.

Table 1.2 1st © Paketesama/stock.adobe.com, 2nd © Revenaif/Shutterstock.com; Table 1.3 Images courtesy of Draper Tools Ltd www.drapertools.com; Table 1.4 1st © Modustollens/stock.adobe.com, 2nd © Screwfix Direct Limited, 3rd © Vladimir Liverts/stock.adobe.com, 4th © Screwfix Direct Limited, 5th Images courtesy of Draper Tools Ltd www.drapertools.com; Table 1.5 1st © Vvoe/stock.adobe.om, 2nd © Dp3010/stock.adobe.com, 3rd © Aldorado/stock.adobe.com, 4th and 5th Images courtesy of Draper Tools Ltd www.drapertools.com; Table 1.6 1st © Lunglee/stock.adobe.com, 2nd © Molnia/stock.adobe.com; Table 1.7 Images courtesy of Draper Tools Ltd www.drapertools.com; Table 1.8 1st Images courtesy of Draper Tools Ltd www.drapertools.com, 2nd © Michaklootwijk/stock.adobe.com, 3rd © Dmitriy Syechin/stock.adobe.com; Table 1.9 1st © Vj/stock.adobe.com, 2nd © Remedia/stock.adobe.com, 3rd © Screwfix Direct Limited, 4th © Artburger/stock.adobe.com, 5th Images courtesy of Draper Tools Ltd www.drapertools.com; Table 1.10 1st © Alexstar/stock.adobe.com, 2nd © Maxximmm/stock.adobe.com, 3rd © Sergey Sosnitsky/stock.adobe.com, 4th © Cristi180884/stock.adobe.com; Table 1.11 1st © David J. Green/Alamy Stock Photo, 2nd © Rapheephat/stock.adobe.com; Table 1.12 1st © Metabo, 2nd and 3rd © Screwfix Direct Limited; Table 1.13 1st © Bradcalkins/stock.adobe.com, 2nd © Anton/stock.adobe.com, 3rd © Vladimir Zubkov/stock.adobe.com, 4th and 5th © Screwfix Direct Limited, 6th © Luckylight/stock.adobe.com; Table 1.14 1st © Stoleg/stock.adobe.com, 2nd © Roman Milert/stock.adobe.com, 3rd © Eugene Shatilo/stock.adobe.com; Table 1.15 1st Images Courtesy of RIDGID® - RIDGID® is the registered trademark of RIDGID, Inc., 2nd Photograph by kind permission of ROTHENBERGER UK Ltd; Table 1.16 © ROTHENBERGER; Table 1.17 Images Courtesy of RIDGID® - RIDGID® is the registered trademark of RIDGID, Inc.; Table 1.18 Photograph by kind permission of ROTHENBERGER UK Ltd; Table 1.20 1st © Astroflame Fireseals Ltd, 2nd Reproduced by kind permission of HSE, HSE would like to make it clear it has not reviewed this product and does not endorse the business activity of Hodder Education; Table 1.21 1st © Roadknight/stock.adobe.com, 2nd © Remus20/stock.adobe.com, 3rd © Cvetanovski/stock.adobe.com, 4th and 5th © Screwfix Direct Limited, 6th © Images reproduced by kind permission of Rainclear Systems Ltd. UK; Table 1.22 1st © Trading Depot, 2nd © Toolstation Ltd, 3rd and 4th © Toolstation Ltd, 5th © Screwfix Direct Limited, 6th © Toolstation Ltd, 7th © GF Piping Systems, 8th © Aviavlad/stock.adobe.com; Tables 1.24, 1.25, 1.26, 1.27 and 1.28 © Pegler Yorkshire Group; Table 1.29 1st © Toolstation Ltd, 2nd and 4th © Pegler Yorkshire Group; Table 1.30 1st © Pegler Yorkshire Group, 2nd © Philmac; Table 1.32 1st–4th, 6th, 7th, 9th © Toolstation Ltd, 5th © LisAnn/stock.adobe.com, 8th and 10th © Images courtesy of drainageonline.co.uk; Table 1.33 1st, 2nd, 4th and 6th © Images supplied by Polypipe Building Products, 3rd © MTG/stock.adobe.com, 5th © Toolstation Ltd; Table 1.34 © Toolstation Ltd; Table 1.35 1st © Wavin Limited, 2nd © John Guest, 3rd © Trading Depot; Table 1.38 1st © Arbalest/stock.adobe.com, 2nd © Dionisvera/stock.adobe.com, 3rd and 6th © Amnach/stock.adobe.com, 4th © Unkas Photo/stock.adobe.com, 5th © Cegli/stock.adobe.com, 7th © Sompob wongnuksue/123RF; page 30, Q9 © Bradcalkins/stock.adobe.com; Table 2.17 1st, 2nd and 4th© Contactum Limited, 3rd © Heating Parts Specialists Ltd; Table 2.18 © Ultimatehandyman.co.uk; page 98 © Purmo Group Ltd; page 120 left © Juan/stock.adobe.com, right © JE-MTY/Shutterstock.com; page 150 © Jusep/stock.adobe.com; page 161 Q25 © Jusep/stock.adobe.com.